Springer Proceedings in Physics

Volume 209

The series Springer Proceedings in Physics, founded in 1984, is devoted to timely reports of state-of-the-art developments in physics and related sciences. Typically based on material presented at conferences, workshops and similar scientific meetings, volumes published in this series will constitute a comprehensive up-to-date source of reference on a field or subfield of relevance in contemporary physics. Proposals must include the following:

- name, place and date of the scientific meeting
- a link to the committees (local organization, international advisors etc.)
- scientific description of the meeting
- list of invited/plenary speakers
- an estimate of the planned proceedings book parameters (number of pages/ articles, requested number of bulk copies, submission deadline).

More information about this series at http://www.springer.com/series/361

H. Paul Urbach · Qifeng Yu
Editors

4th International Symposium of Space Optical Instruments and Applications

Delft, The Netherlands, October 16–18, 2017

 Springer

Editors
H. Paul Urbach
Optics Research Group
Delft University of Technology
Delft, The Netherlands

Qifeng Yu
College of Aerospace Science/Engineering
National University of Defense Technology
Changsha, China

ISSN 0930-8989   ISSN 1867-4941   (electronic)
Springer Proceedings in Physics
ISBN 978-3-030-07235-3   ISBN 978-3-319-96707-3   (eBook)
https://doi.org/10.1007/978-3-319-96707-3

This Springer imprint is published by the registered company Springer Nature Switzerland AG
The registered company address is: Gewerbestrasse 11, 6330 Cham, Switzerland

Introduction

October 16–18, 2017, the 4th International Symposium on Space Optical Instruments and Applications was successfully held at Delft, the Netherlands. The goal of this yearly symposium is to encourage international communication and cooperation in space optics and promote the innovation, research, and engineering development, initiated by the Sino-Holland Space Optical Instruments Joint Laboratory. The symposium focuses on key innovations of space-based optical instruments and applications, and the newest developments in theory, technology, and applications in optics in both China and Europe. It will provide an opportune moment for exchanges on space optical area and information on current and planned optical missions.

The speakers include distinguished guests from ESA, CNSA, NSO as well as renowned European and Chinese universities, space institutes and companies. There were nearly 60 presentations given by speakers from about 40 Chinese and European government sectors, enterprises, organizations, and universities. The total number of participants reached 150.

The main topics include
 Space optical remote sensing system design;
 Advanced optical system design and manufacturing;
 Remote sensor calibration and measurement;
 Remote sensing data processing and information retrieval;
 Remote sensing data applications.

Organization

Hosted by

Jointly Hosted by

Organized by

Jointly Organized by

Chairmen

Paul Urbach	TU Delft
Yu Qifeng	NUDT

Executive Chairmen

Kees Buijsrogge	TNO
Hu Chen	BISME

Secretary-General

Andrew Court	TNO
Bin Fan	BISME

Organizing Committee

Bart Snijders	TNO
Jing Zou	TNO
Yue Li	BISME
Paul Urbach	TU Delft
Yvonne van Aalst	TU Delft
Riet Schutz	TNO
Sandra Baak	TNO

Contents

Optical Design and Simulation of Spatial Heterodyne Spectroscopy (SHS) for Mesospheric Temperature

Haiyan Luo[✉], Xuejing Fang, Wei Xiong, and Hailiang Shi

Key Laboratory of Optical Calibration and Characterization,
Anhui Institute of Optics and Fine Mechanics,
Chinese Academy of Sciences, Hefei, China
luohaiyan@aiofm.ac.cn

Abstract. The mesospheric region exists a lot of complex photochemistry reactions and dynamical processes, which plays an important role in affecting the dynamics in global circulation models and the safety for aircrafts. And gravity wave is a significant dynamical factor in thermodynamic structure and circulation structure. However, because of rare atmosphere in this region, the high-precision measurements on features of mesospheric temperature are insufficient due to a lack of measurement techniques. Design and performance parameters for a high spectral resolution, high spatial resolution in vertical direction, robust and CubeSat-scale spatial heterodyne spectrometer for mesospheric temperature detection are reported. According to the measured value of in-orbit payloads, the subsystem parameters of instrument are described in detail by using spatial heterodyne spectroscopy (SHS) in the O_2 atmospheric-band (A-band) (0-0). By optical design and simulation for the instrument with high spectral resolution and spatial resolution in one dimension, we can get the quantitative relationship between the spectral stability and the retrieved temperature precision. For 0.5% relative spectral intensity stability, it has a theoretical temperature precision of 2 K by fitting three peaks, and the signal to noise ratio will be improved with a higher spectral resolution and more emission spectral peaks under the same equivalent noise condition. Finally, orbit attitude control system requirements can be estimated by considering orbit height and the instrument parameters. The scan is not accomplished through controlled nodding or rotation of the spacecraft during short exposure time, and the spectral background signals due to narrow Fraunhofer features and O_2-A band absorption lines below 90 km is detectable at SHS spectral resolution. These results will provide a new method to get the global distribution of airglow in wide altitude range (40–130 km) information with simultaneously imaging technique for each view slice of the scene. It will provide basic data for studying the mesospheric temperature and gravity wave activity in the mesospheric atmosphere circulation models.

Keywords: Spatial heterodyne spectroscopy (SHS) · Mesospheric temperature Oxygen airglow spectra

© Springer Nature Switzerland AG 2018
H. P. Urbach and Q. Yu (Eds.): ISSOIA 2017, SPPHY 209, pp. 1–10, 2018.
https://doi.org/10.1007/978-3-319-96707-3_1

1   Introduction

The MEsospheric temperature observations of airglow spatial heterodyne Imaging Instrument (MEII) is a spatial heterodyne spectrometer which measures limb airglow emission from the oxygen infrared atmospheric band near 762 nm. Studying relative spectral intensity of airglow emission and deriving temperature vertical profile and gravity wave activity in the mesosphere and low thermosphere (MLT) region of the atmosphere are critical to understanding their impact on climate. In addition, many atmospheric chemical reactions in the MLT region have significant temperature dependence [1, 2].

Temperature vertical profiles can be derived from LIDAR and other ground-based instruments, but they are insufficient to understand the region globally due to the limited geographic and temporal coverage. The global information can be achieved with optical payloads for space or satellite missions. The Optical Spectrograph and InfraRed Imaging System (OSIRIS) on board the Odin satellite is a limb-viewing instrument that consists of an Optical Spectrograph (OS) and an InfraRed Imager (IRI) [3]. Currently, the OSIRIS instrument retrieves MLT temperatures from observations of the O_2-A band dayglow emission spectrum, with an estimated accuracy of $\pm$ 2 K near 90 km and less than $\pm$ 6 K near 105 km with exposure time more than 2 s and vertical spatial resolution of 5 km [4]. However, the spectral signals in the OS observation, due to narrow Fraunhofer features and O_2-A band absorption lines, is essentially undetectable at OS spectral resolution, and the scan is accomplished through controlled nodding or rotation of the spacecraft during long exposure time.

The SHS technology has combined characteristics of a Fourier Transform Spectrometer and a Fabry-Perot interferometer with a high spectral resolution and narrow spectral range. Furthermore, it has already been successfully tested in a space environment [5, 6]. The current concept allows sampling specific airglow spectrum of O_2-A band across a large range of altitude in a single exposure by SHS. In this paper, MEII designs for observation of O_2-A band spectrum are discussed. As an instrument within a small volume and without moving parts, MEII has high potential to increase the importance of CubeSat for science missions.

2   Oxygen Airglow Emission Feature and MEII Requirements

The mesosphere is too thin to be detected directly with LIDAR and other ground-based instruments, and it is also weakened due to absorption of the airglow emission below 90 km. Fortunately, however, many thermally excited molecular and atomic species in MLT region can be detected by remote sensor with limb-viewing. The O_2-A airglow features are one of the strongest band in all candidate mesospheric emission features, and the radiative lifetime is one of the longest band lasting for 12 s, so that the retrieved temperature is the MLT region temperature. Spectral radiance from a typical OS limb scan is shown in Fig. 1, which was created by using different tangent height measurements at different time over a period of around 2 min [3]. We can see that the O_2-A band (758–771 nm) spectral features appear as emission features at high tangent height and strong absorption features at low tangent height as shown in Fig. 1. Thus, limb observation are assumed, and the emission features and background features are considered from Fig. 1 at different tangent heights: the airglow emission intensity near 90 km (3.6e11 photons $cm^{-2} s^{-1} nm^{-1} sr^{-1}$), the background intensity near 90 km (6.0e9 photons $cm^{-2} s^{-1} nm^{-1} sr^{-1}$). The OS of OSIRIS specifications and basic parameters given in Table 1.

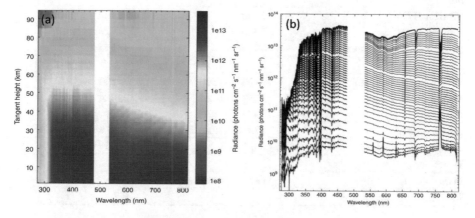

Fig. 1. Spectral radiance from a typical OS of OSIRIS limb scan

The associated signal-to-noise ratio (SNR) can be estimated near 90 km in the paper by the MEII instrument with the measurement radiance of OS, which accounts for the photon shot noise from the airglow at the detector, the shot noise because of the detector dark current (3.3 electrons), the readout noise (9.88 electrons) and quantization noise (14.1 electrons). The MEII specifications and basic parameters are also given in Table 1 under specific SNR. The MEII operates with SHS and one-dimensional imaging technology, where the limb-scatter measurements from the different tangent height are detected at one exposure time. So that, it is not necessary to combine with the nodding motion of the spacecraft, the vertical spatial resolution of the limb-scatter measurements lies on the recovered temperature accuracy and the associated signal-to-noise ratio.

In fact, the spectral resolution is broadened as due to the instrument line of shape (ILS) for each emission line or monochromatic spectrum. The right side of Fig. 2 shows the results of analysis for a spectral resolution of 0.047 nm, and the double emission lines will be mixed if the spectral resolution or ILS is more than 0.06 nm. Theoretical studies have indicated that multi-number emission lines and high spectral resolution can effectively increase SNR. As shown in Fig. 3, relative SNR is dependent on the spectral resolution with assumed SNR of 1000 at spectral resolution of 0.001 nm, and enlarged detail relationship with the spectral resolution varied from 0.03 nm to 0.12 nm is shown on the right side of Fig. 3.

Table 1. The OS of OSIRIS specifications and MEII requirements for the O_2-A band

	OS	MEII
Spectral resolution/range	~1.2 nm/759–767 nm	~0.04 nm/762–764.6 nm
Vertical spatial resolution (without binning)	1 km	0.6 km
Vertical IFOV	1 km	90 km
Vertical FOV	5 km	2.5 km
Horizontal spatial resolution	40 km	185 km
Exposure time	2 s @ height of 90 km	1 s @ height of 90 km
SNR (background, emission)	~200, ~500	~100, ~560
Detector parameters	1353 × 143 @ 20 × 27 µm	720 × 320 @ 6.5 × 6.5 µm
Readout noise	17.6 electrons RMS	17.5 electrons RMS
Temperature stability	±0.1 °C	±1.5 °C
F-number	/	f/4
Instrument technology	Grating spectrograph	SHS
Interferogram modulation	/	0.7

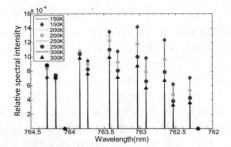 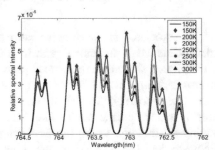

Fig. 2. Relative emission intensity with spectral resolution of 0.001 nm and 0.047 nm for four temperatures.

3   1-D Imaging Spectrometer

High SNR needs several spectral lines with specific spectral resolution to be measure to recover the MLT temperature. A spatial heterodyne spectrometer can measure several spectral lines simultaneously with ultra-spectral resolution. The SHS was developed since 1990s by Harlander and co-authors and is also regularly produced at this time [5–8]. In the following, the concept of SHS and one-dimensional imaging technology with the limb-observation mode are explained.

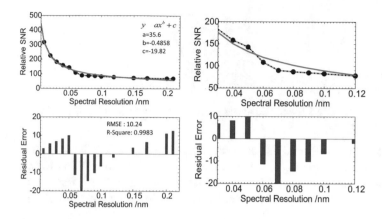

Fig. 3. (a) Relative SNR (Signal to noise) in dependence of the spectral resolution with assumed SNR of 1000 at spectral resolution of 0.001 nm. (b) Enlarged detail of (a) with spectral resolution of 0.03–0.12 nm.

3.1 Field Widened Spatial Heterodyne Interferometry (FW-SHS)

FW-SHS in concept is similar to the conventional field widened FTS system, by insertion of prisms with appropriately wedge angles and material (refractive index: n) so that the optical path difference generated by off-axis beams is compensated. For an input spectral intensity $B(\sigma)$ with an idea point source, the interferogram intensity $I(x)$ recorded as a function of position x along the dispersion dimension is given by

$$I(x) = B(\sigma)\{1 + \cos[2\pi(\sigma - \sigma_0)4x\tan\theta]\} \tag{1}$$

where σ_0 is the Littrow wavenumber of the gratings, θ is the Littrow blaze angle of the gratings, and x is measured in the localization plane of the gratings.

Comparing to SHS with an idea point source, field widened SHS is shown in Fig. 4 where the filed widened prism is placed in each arms of interferometer. For a solid angle Ω, the phase of the cosine interferogram $\delta_{\Omega,\sigma}$ is recast as (2).

$$\delta_{\Omega,\sigma} = \delta_\sigma + 2\pi x\left[2\beta^2\left(\frac{2(n^2-1)\tan\gamma}{n^2\cos^2\frac{\alpha}{2}} - \tan\theta\right) + 2\phi^2\left(\frac{(n^2-1)(2n^2-\sin^2\gamma)\tan\gamma}{n^2(n^2-\sin^2\gamma)} - \tan\theta\right) + \cdots\right] \tag{2}$$

where angle β is between the incident ray and the optical axis, angle ϕ is between the incident ray and the dispersion plane, n is the refractive index of the prisms, γ is the incident angle of the prisms, and α is the wedge angle. With the help of (2), the interferogram intensity $I_\Omega(x)$ can be written as (3).

$$I_\Omega(x) = kB(\sigma)\{1 + \sin c(\frac{u\sigma\Omega C(n,\gamma)}{2\pi\tan\theta})\cos[2\pi u(\sigma(1 - \frac{\Omega C(n,\gamma)}{2\pi\tan\theta}) - \sigma_0)]\} \tag{3}$$

If the maximum solid angle is used, FWHM of the previous instrumental line shape $(1.207\delta\sigma)$ will be slightly increased and the interferogram modulation efficiency will be

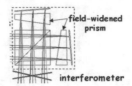

Fig. 4. Schematic diagram of field widened SHS.

dropped to $\sin c(1/2) = 0.637$ at the maximum optical path difference point. However, the field widened prism also produce a small resolution in addition to interferometer, and it will increase about 2% for the prisms with negative refractive index. These differences should be taken into account when calibrating the spectral location.

3.2 One-Dimensional Imaging Fore-Optics and Imaging Optics

The cylindrical lens is used to image the measured target in one dimension at the detector, and the cylindrical lens can be located on the front of the collimating lens or inside of the imaging lens. The collimator was used together with the cylindrical lens for imaging the along-spatial targets and illuminating the along-spectral targets as shown in Fig. 5. The imaging lens re-images the fringes located close to the gratings onto the 2-D detector, and each spectral information of different height in MLT can be expressed as a spatial slice imaged onto the detector array. The MEII optical system is composed of a cylindrical lens, collimating lens, field widened spatial heterodyne interferometry, imaging lens and detector.

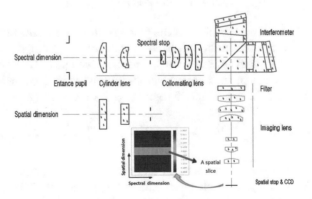

Fig. 5. Optical layout of MEII.

For a 2-D imaging version of the interferometer, the image of the scene is scanned across the fringe and the Fourier transform is applied to synthesized interferogram formed a sequence of frames and representing the fringe pattern multiplied by individual spatial elements of the target scene. For a 1-D imaging version of MEII, the image of the scene is obtained at same exposure time without scanning.

The detector array made by e2v technologies has an active imaging area of 1024 by 1024 pixels with a size of 6.5 by 6.5 μm pixel pitch, so that the 720 pixels are in the dispersion direction and the 320 pixels in the spatial direction. The imaging lens has a magnification of 2.38 × with a 13.76 mm grating along the dispersion direction. The aberration properties of the MEII subsystem are accurately modeled as shown in Table 2. The modulation effect factors of SHS by the extended source should be more than 0.86, and here, the modulation effect factor of field widened spatial heterodyne interferometry is 0.89.

Table 2. The aberration properties of the subsystem

Collimating lens	Effective focal length	173 mm
	F-number	9.5
Imaging lens	magnification	−2.38:1(Grating/Detector)
Beam Splitter	Size	25.4 × 25.4 mm
Grating	Groove density	600gr/mm
	Clear aperture	13.76 × 5.80 mm
	Littrow angle	13.32°
Interferometry	Temperature Stability	±1.5 °C
	Littrow wavelength	768 nm
Field widening prisms	Substrate	N-SF57

4 Inversion Method and Simulation

The MEII views the limb of the day airglow and so the emission rate measured in each bin is the line of sight integral of the volume emission rate modified by SHS as shown in Fig. 6 and (4). It is also important to note that spacecraft had a sun-facing side and a dark side, and MEII was mounted on the sun-facing side.

$$O_i = 2\sum_{j=i}^{N} V_i L_{i,j}, \quad i = 1 \sim N \tag{4}$$

where $L_{i,j}$ is the path length of ith observation line of sight through the jth atmospheric height, and the viewing geometry is shown in Fig. 7. In fact, the absorption of the airglow emission should be considered below 70 km by multiplying L_{ij} with corresponding absorption factor in the inversion process.

$$O_N = 2L_{N,N}V_N \quad \Rightarrow V_N = O_N/2L_{N,N} \tag{5}$$

$$L_{N,N} \approx \sqrt{[R + H_0 + N\delta]^2 - [R + H_0 + (N-1)\delta]^2} \tag{6}$$

where $L_{N,N}$ is the path length of Nth observation line of sight through the highest atmosphere by MEII, and atmospheric refraction effect is ignored in (6).

$$S(z, T) = \sum_{\lambda_1}^{\lambda_N} |V_\lambda(z) - E_\lambda(T)|^2 \tag{7}$$

A nonlinear least squares algorithm was used to adjust the theoretical emission spectra $E(T)$ to fit the inverted spectra V_i using space-based observation spectra O_i as shown in (7) and (4).

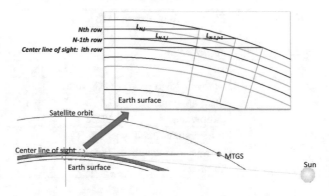

Fig. 6. Illustration of limb measurement concept and the light of sight geometry

Typical ideal spectra for O_2 A-band for four mesopause temperature value are shown in Fig. 8, and each emission spectra from No. 4 to No. 8 indicate significant modulation for different temperature. By the spectral intensity of No. 5, No. 6 and No. 7, we can get the coefficients of quadratic polynomial as shown in Table 3 and (8).

$$y = a + b_1 x + b_2 x^2 \tag{8}$$

For 0.5% relative intensity stability of each emission spectral line, it has a theoretical temperature precision of 2 K by fitting three peaks, and the signal to noise ratio will be improved with a higher spectral resolution and more emission spectral peaks under the same equivalent noise condition.

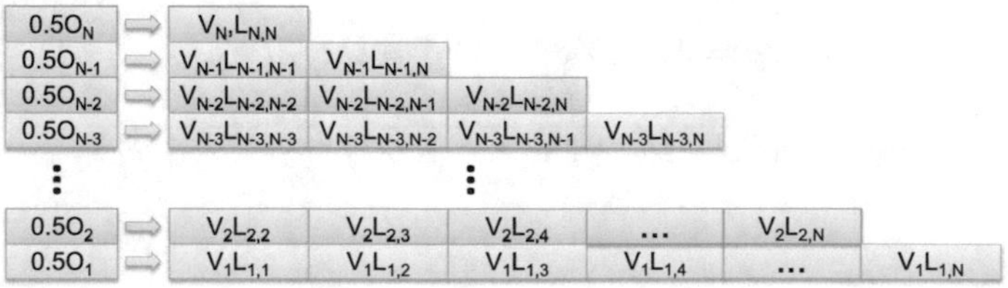

Fig. 7. Viewing geometry in (4)

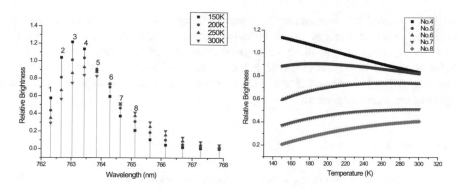

Fig. 8. Typical O_2 A-band (0-0) spectra for different temperature

Table 3. Polynomial coefficients with different temperature and relative radiance

Polynomial coefficients	Different temperature			Different relative radiance		
	198 K	200 K	202 K	+0.5% @200 K	200 K	−0.5% @200 K
b_1	−0.35696	−0.34537	−0.33396	−0.35775	−0.34537	−0.33295
b_2	−0.26723	−0.27699	−0.28653	−0.27229	−0.27699	−0.28174
a	0.90613	0.90554	0.90488	0.91007	0.90554	0.90102

5 Conclusion

A theoretical optical design for a limb-sounding payload MEII is presented according to the absolute radiance of OSIRIS in orbit, and MEII provides spatial resolution of order 600 m with a swath width 185 km in a sun-synchronous orbit of about 895 km altitude. MEII is mounted on satellite such that the sensor optical axis oriented towards limb. Attitude control system requirements can be estimated by considering the field of view for one exposure time (1 s), 2.5 km × 185 km, relative to an assumed working distance of 3300 km above the O_2 airglow emission layer. Yaw precision of at least $0.77° = \tan^{-1} (2.5 \text{ km}/185 \text{ km})$ and pitch precision of $0.04° = \tan^{-1} (2.5 \text{ km}/3300 \text{ km})$ should be maintained, and orientation precision can be coarser, as its effects are to harmlessly bias the field of view in vertical direction.

For 0.5% relative spectral stability, using a quadratic algorithm by three peaks (764.2804 nm, 764.6296 nm and 765.1132 nm) in O_2 atmospheric-band (A-band) (0-0), MEII can has a theoretical temperature precision of 2 K with exposure time <1 s and temperature controller accuracy of 1.5 °C. Different from measuring wind payloads by SHS, MEII has an approximate two order of magnitude improvement in temperature control accuracy, and the MEII spectrum can be shifted and stretched in orbit if it is necessary.

The MEII SNR and spectrum quality are also directly influenced by interferogram modulation efficiency. This paper present a research work on the relations between modulation efficiency and the extended source, and the decrease of modulation efficiency caused by performance of beam splitting coating, optical aberrations and pixel size of detector will be discussed in our future work. For a 1-D imaging MEII in nadir mode, small and medium scale gravity wave's features can be also detected. Without any moving parts and with compact size and a relative loose thermal environment, MEII is well suited for the observation of the trace elements on future CubeSat for science missions.

References

1. Doe Richard, A., Watchorn, S., Butler, J.J., Xiong, X., Gu, X.: Climate-monitoring CubeSat mission CM2: a project for global mesopause temperature sensing. SPIE Opt. Eng. Appl. **8153**, 81530Q-1–81530Q-10 (2011)
2. Beres, J.H., Carcia, R.R., Boville, B.A., Sassi, F.: Implementation of a gravity wave source spectrum parameterization dependent on the properties of convection in the Whole Atmosphere Community Climate Model (WACCM). J. Geophys. Res. **110**, 10108 (2005)
3. Llewellyn, E.J., Lloyd, N.D., Degenstein, D.A., Gattinger, R.L., Petelina, S.V., Bourassa, A. E., Wiensz, J.T., Ivanov, E.V., McDade, I.C., Solheim, B.H., McConnell, J.C., Haley, C.S., von Savigny, C., Sioris, C.E., McLinden, C.A., Evans, W.F.J., Puckrin, E., Strong, K., Wehrle, V., Hum, R.H., Kendall, D.J.W., Matsushita, J., Murtagh, D.P., Brohede, S., Stegman, J., Witt, G., Garnes, G., Payne, W.F., Piché, L., Smith, K., Warshaw, G., Deslauniers, D.L., Marchand, P., Richardson, E.H., King, R.A., Wever, I., McCreath, W., Kyrölä, E., Oikarinen, L., Leppelmeier, G.W., Auvinen, H., Mégie, G., Hauchecorne, A., Lefèvre, F., de La Nöe, J., Ricaud, P., Frisk, U., Sjoberg, F., von Schéele, F., Nordh, L.: The OSIRIS instrument on the Odin spacecraft. Can. J. Phys. **82**(6), 411–422 (2004)
4. Sheese, P.E., Llewllyn, E.J., Gattinger, R.L., Bourassa, A.E., Degenstein, D.A., Lloyd, N.D., McDade, I.C.: Temperature in the upper mesosphere and lower thermosphere from OSIRIS observation of O_2 A-band emission spectra. Can. J. Phys. **88**, 919–925 (2010)
5. Harlander, J.M., Roesler, F.L., Cardon, J.G., Englert, C.R., Conway, R.R.: SHIMMER: a spatial heterodyne spectrometer for remote sensing of Earth's middle atmosphere. Appl. Opt. **41**(7), 1343–1352 (2002)
6. Englert, C.R., Stevens, M.H., Siskind, D.E., Harlander, J.M., Roesler, F.L.: Spatial heterodyne imager for mesospheric radicals on STPSat-1. J. Geophys. Res. **115**, D20306 (2010). https://doi.org/10.1029/2010JD014398
7. Luo, H., Xiong, W., Li, S., Li, Z., Hong, J.: Optical design and measurements of a dynamic target monitoring spectrometer for potassium spectra detection in a Flame. In: 3rd International Symposium of Space Optical Instruments and Applications. Springer Proceedings in Physics, vol. 192, pp. 61–72 (2017)
8. Harlander, J.M.: Spatial heterodyne spectroscopy: Interferometric performance at any wavelength without scanning. The University of Wisconsin – Madison (1991)

System Design Technology of Visual and Infrared Multispectral Sensor

Yanhua Zhao[✉], Liqun Dai, Shaojun Bai, Jianfeng Liu,
and Honggang Peng

Key Laboratory for Advanced Optical Remote Sensing Technology of Beijing,
Beijing Institute of Space Mechanics and Electricity, Beijing, China
zhaoyh304@sina.com

Abstract. In this paper, we introduce the index and scheme of the full spectrum spectrometer on the GF-5, which has the characteristics of wide spectral range, high spatial resolution and high radiometric calibration accuracy. Compared with the similar loads at home and abroad, the technical indicators have reached the leading domestic and international advanced level of technology, so that our country in the spatial resolution of multispectral imaging and high precision observation capabilities have been greatly improved. In this paper, the advanced technology of the load is analyzed.

Keywords: ISSOIA2017 · Hyperspectral · Remote-sensing · GF-5
Visual and infrared multispectral sensor

1 Introduction

GF-5 satellite, which is used to obtain high spectral resolution remote sensing data products from ultraviolet to long wave infrared spectrum. It is a high-sporadic "high-space resolution" Resolution, high spectral resolution and high-precision observation of time and space coordination, all-weather, all-day earth observation system "an important part of the target is to achieve national high-resolution earth observation capabilities one of the important signs.

The Visual and infrared multispectral sensor is one of the main loads of the GF-5 satellite. It is the second generation of the multi-spectral imager from the visible to the hot infrared spectrum. (0.45 μm–12.5 μm) 12-band remote sensing data products, especially the long-wave infrared spectrum can be used to invert the land surface of the remote sensing data, Temperature, water temperature, with night detection capability, at the same time through the split window section, can improve the detection accuracy, thereby enhancing China's environmental monitoring and resource survey and other aspects of the capacity, it is the first to achieve coverage, near infrared, Medium wave, long wave, very long wave infrared spectrum of multi-spectral camera, thermal infrared resolution of 40 m in the international leading level.

Full spectrum of spectral imager with the ability to observe the whole day, can be continuously imaging 6 tracks, each track to the ground imaging 25 min, covering about 12,000 km.

© Springer Nature Switzerland AG 2018
H. P. Urbach and Q. Yu (Eds.): ISSOIA 2017, SPPHY 209, pp. 11–17, 2018.
https://doi.org/10.1007/978-3-319-96707-3_2

2   Development Characteristics and Difficulties Analysis

(1) System indicators require high, domestic development for the first time.

System complex: the use of optical system sub-field, in the convergence of the optical path in the use of color separation, the focus on the use of micro-combination filter spectroscopy and other measures, the system focal length (3 focal plane, 12 spectrum), Standard methods and methods (all-optical path\full-diameter diffuse reflector calibration and partial optical path full-caliber black body calibration), high reliability requirements, the system is very complex;

High spatial resolution: long-wave infrared spectral segment spatial resolution of 40 m, is the current domestic long-wave infrared maximum resolution multi-spectral camera, compared with similar foreign load is also in the leading level;

Radiation resolution is high: long wave channel for the first time using split window detection, high temperature inversion accuracy, high resolution requirements of radiation;

Spectral coverage is wide: from visible to hot infrared (0.45 μm–12.5 μm), long wave cutoff frequency close to very long wave, is the domestic high-resolution multi-spectral remote sensing spectral range of the widest load.

(2) The system uses infrared and visible light common optical path of the new optical system, the optical mirror processing index and the high precision of the installation, optical processing and decoration difficult.

(3) The cooling structure of the chiller is complex and the cooling capacity is large.

3   Overall Program

The overall scheme of the Visual and infrared multispectral sensor is as follows:

(1) The use of large field of view small F number off axis three anti-optical system;

(2) Color separation film combined filter to achieve spectral subdivision;

(3) Imaging method: long line array multi-spectrum push sweep;

(4) Using Multi-spectral TDI CCD, shortwave- medium wave infrared detector, long wave infrared detector, Using TDI is conducive to the consistency of the detection channel output, improve the image quality;

(5) The use of full-caliber black body + diffuse reflector to achieve high-precision calibration, driven by the gear drive diffuse reflector board into the optical path, the introduction of sunlight on the visible near infrared, short-wave infrared spectral section calibration; black body calibration mechanism by step motor drive Respectively, to the temperature of the black body and the self-reflection, the infrared wave, long-wave infrared calibration, and monitor the stability of the detector output; in the launch, the use of electromagnets to lock;

(6) Chiller selection of 60 K and 80 K pulse tube refrigerator, with the cylinder dual piston opposed compressor, can effectively reduce the vibration of the refrigerator and spring movement load.

The working principle is as follows: The Visual and infrared multispectral sensor uses off-axis three anti-main optical system, shortwave, medium wave and long wave infrared spectrum sharing main, secondary and three mirror, after the three mirror to join the color separation film, respectively, Wave, long wave infrared detector on the focal plane; using the field of view separation technology will be visible near the infrared spectrum separated, converged to the CCD detector. Using long-line array TDI detector push method to obtain high-quality multi-spectral earth image, and then by the information acquisition circuit to detect the detector image signal, the AD conversion, sent to the signal processing box in CCSDS format, through the LVDS interface Sent to the stars on the number of sub-system. Using diffuse reflection and blackbody to achieve high precision in-orbit radiation calibration with different spectral bands.

4 Domestic and International Benchmarking

Full spectrum imager (Fig. 1) the spectral range is 0.45 μm–12.5 μm, spatial resolution in the visible and near infrared and shortwave infrared spectrum 20 m in the infrared spectrum of 40 m.

Fig. 1. Visual and infrared multispectral sensor physical map

According to public information query, considering spectrum setting, spatial resolution, the width and the direction of application and so on, can the analogy of similar domestic and foreign load has the United States Landsat-7 ETM+, Landsat-8 OLI/TIRS, MTI in the United States and China Resources star No. 04 infrared camera.

Table 1 Comparison of indicators. By comparison, we can see that in the same kind of load, the full spectrum imager has the highest spectrum and the highest resolution.

Table 1. Domestic and foreign related load technical index

The main indicators	Remote sensor				
	ETM+ (landsat-7, America), the launch of 1999.4.15, 2005 retired	OLI+ TIRS (landsat-8, America), launch of 2013.2.11	MTI	IRS (ZY-1 04, China), launch of 2014.12.7	VIMS (GF-5, China), expected to launch of 2017.9
Bands (µm)	0.52–0.9, 0.45–0.53, 0.52–0.60, 0.63–0.69, 0.76–0.90, 1.55–1.75, 2.09–2.35, 10.4–12.5	OLI: 0.5–0.68, 0.43–0.45, 0.45–0.51, 0.53–0.59, 0.64–0.67, 0.85–0.88, 1.36–1.38, 1.57–1.65, 2.11–2.29 TIRS: 10.60–11.19, 11.50–12.51	0.45–0.52, 0.52–0.60, 0.62–0.68, 0.76–0.86, 0.86–0.90, 0.91–0.97, 0.99–1.04, 1.36–1.39, 1.55–1.75, 3.49–4.10, 4.85–5.05, 8.01–8.39, 8.42–8.83, 10.15–10.70, 2.08–2.35	0.50–0.90, 1.55–1.75, 2.08–2.35, 10.4–12.5	0.45–0.52, 0.52–0.60, 0.62–0.68, 0.76–0.86, 1.55–1.75, 2.08–2.35, 3.50–3.90, 4.85–5.05, 8.01–8.39, 8.42–8.83, 10.3–11.3, 11.4–12.5
The number of spectral bands	8	11	15	4	12
Orbit	705 km	705 km	575 km	778 km	705 km
Sub pixel resolution	P: 15 VNIR: 30 m SWIR: 30 m TIR: 60 m	P: 15 VNIR: 30 m SWIR: 30 m TIR: 100 m	VNIR: 5 m, other: 23 m	VIS/SWIR: 40 m, TIR: 80 m	VIS/NIR/SWIR: 20 m, MWIR/LWIR : 40 m
Coverage width	185 km	185 km	12 × 12 km	120 km	60 km
The absolute radiation measurement uncertainty	5%	5%	1 3%	10%	5%, 1 K
Design life	5 years	5 years	3 years	3 years	8 years

5 Advanced Technology Analysis

5.1 Spectral Range Wide, High Spatial Resolution

(1) Technical Features

Full spectrum imager covering 0.45 µm–12.5 µm a total of 12 bands, including the visible, near infrared and shortwave, medium wave, long wave spectrum, in the similar load spectrum in the most visible and short wave infrared spectroscopy; multi spectral spatial resolution of 20 m in infrared spectrum the spatial resolution of 40 m, is currently the highest international.

(2) Means of Realization

According to the spectral width, spectrum, medium focal length, large relative aperture, the transfer function of high requirements, visible infrared optical path and other difficulties, the design of common path two imaging three mirror anastigmatic reflective structure (Fig. 2), using the off-axis aperture to eliminate the secondary mirror and three mirror center masking by folding mirror visible light field separation imaging in multicolor CCD, with color transflective characteristics of the short medium and long wave imaging respectively in two line array infrared detector, greatly improve the system integration degree.

Fig. 2. Optical design

In order to solve the problem of 60 km width, short medium and long wave with long linear HgCdTe TDI type infrared detector, the detector chip size limit, the three module form long implementation.

In order to meet the deadline, the high transmittance of narrow band filter, wide range, high requirement of steepness, taking advantage of long wave and short wave pass film, half wave film etc., through the combination of global and local optimization, to achieve optimal design of the narrowband film system.

5.2 High Radiometric Calibration Accuracy

(1) Technical Features For the quantitative application of the Visual and infrared multispectral sensor, the precision of the visible wave absolute radiometric calibration can reach 5%, and the mid wavelength reaches 1 K@300 K.

(2) Means of Realization If the ground calibration field is only used for radiometric calibration, the efficiency is low and the influence factors are many. Therefore, the satellite radiation calibration device is designed, which can be calibrated in time.

In the visible near infrared and short wave spectrum, the diffuse reflection plate is used to measure the full aperture radiation of the diffuse reflection plate, and the performance of the diffuse reflection plate is monitored by the radiometer. Figure 3 is the visible shortwave spectral segment in the laboratory radiometric calibration test.

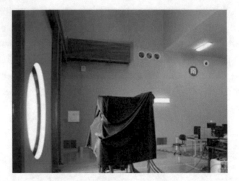

Fig. 3. Visible short wave spectral radiometric calibration test

In the mid long wavelength infrared spectrum, the full aperture radiometric calibration of the partial optical path is realized by using the star shaped blackbody in temperature, as shown in Fig. 4. Before the launch, the laboratory vacuum radiation calibration was performed to obtain the quantitative relationship between the radiance and DN value and the relationship between the internal and external calibration.

Fig. 4. Infrared radiation calibration test

6 Application Direction

Multi spectral imaging remote sensing data has been widely used in resource evaluation, environmental monitoring, disaster early warning and post disaster assessment, urban planning, target classification and recognition, and has great social and economic benefits.

The common optical path of visible and infrared multi spectral imaging, multi spectral image data can obtain high spatial resolution multi spectral coverage of visible and shortwave, medium wave and long wave infrared, spatial matching and spectral matching from the two aspects of analysis and identification of targets can obtain object information more abundant, so to improve the detection ability of the target and the level of emergency monitoring.

The common path of visible and infrared multispectral imaging instrument, can obtain the objects in the visible spectral information of multiple spectral LWIR, especially in the medium and long wave infrared spectrum can be achieved on the surface of the diurnal observation data, and improve the detection accuracy of the split window spectrum, visible and infrared light path is more spectral imager will play an important role in monitoring and resource environment in China survey application.

The Visual and infrared multispectral sensor can be applied to the following aspects:

(1) The field of ecological environment: natural ecological remote sensing monitoring, environmental destruction of large-scale engineering development, urban ecological environment monitoring, rural ecological environment monitoring, biodiversity monitoring;

(2) Land resources: land resources survey, mineral resources exploration and evaluation;

(3) Forestry: desertification monitoring, forest fire monitoring, wetland resources monitoring;

(4) Water resources: water resources management and ecological protection, environmental monitoring and management of rivers and lakes;

(5) Agriculture: agricultural resources survey, crop growth and yield monitoring;

(6) The field of disaster prevention and mitigation: flood disaster monitoring, drought disaster monitoring.

References

1. Tian, G.: Thermal infrared remote sensing, pp. 168–218. Electronic Industry Press, Beijing (2006)
2. Chen, S., Yang, B., Wang, H.: Space Camera Design and Experiment, pp. 24–25, 334–360. China Astronautics Press, Beijing (2003)
3. Liang, S.: Quantitative Remote Sensing of Land Surfaces, pp. 130–142. Science Press, Beijing (2009)
4. Ma, W.: Aerospace Optical Remote Sensing Technology, pp. 210–211. China Science and Technology Press, Beijing (2011)
5. Xu, B.: Meteorological Satellite Payload Technology, pp. 66–68. China Aerospace Press, Beijing (2005)
6. Hang, Q., Pan, Z., Wang, A.: In-orbit radiometric calibration and quantitative application for civil remote sensing satellite payloads. Spacecr. Recovery Remote Sens. 34, 57–65 (2013)
7. Li, X.: Principles and Applications of Remote Sensing, p. 85. Science Press, Beijing (2008)
8. Zhu, L., Gu, X., Chen, L., et al.: Comparison of LST retrieval precision between sigil-channel and split-window for high-resolution infrared camera. J. Infrared Millim. Waves 27(5), 346–353 (2008)

Multi-component Atmosphere Detection Technology Based on Space-Based Filament Laser

X. Liu[1]([⊠]), Y. L. Tao[1], W. W. Liu[2], Z. F. Feng[3], H. Y. Song[4],
S. W. Xu[1], W. Li[1], N. J. Ruan[1], S. B. Liu[4], and Y. C. Zheng[1]

[1] Beijing Institute of Space Mechanics and Electricity, CAST,
Beijing 100094, China
liuxun_laby@163.com
[2] Institute of Modern Optics, Nankai University, Tianjin 300350, China
[3] College of Applied Science, Taiyuan University of Science and Technology,
Taiyuan 030024, China
[4] Institute of Laser Engineering, Beijing University of Technology,
Beijing 100124, China

Abstract. With the increasing severity of environmental problem, the multi-component atmosphere detection has become the research focus. In this paper, the technology based on space-based filament laser is studied in theory, simulation and experiment. When femtosecond laser pulse propagates in air, the natural direction can be restrained, and the filament laser will be generated due to the nonlinear self-focus effect. The diameter of filament laser is about 100 μm, so the laser intensity is about 10^{14} W/cm^2 inside the filaments laser. When the filament laser interacts with the air, the pollutants are dissociated into small fragments, and the molecules and atoms are ionized. The fluorescent fingerprint spectrum will be emitted according to the special chemical composition, which can be applied in detecting the atmospheric components and identifying the pollutants. The analysis of the pollutant concentration can be performed through direct measurements of the characteristic spectral lines. Relying on the possible generation of filaments at dozens of kilometric distances, this technology can be used in remote analysis of hazardous or unreachable spots, which can carry out the real-time vertical dimensional detection in large scale continuously.

Keywords: Filament · Femtosecond laser pulse · Self-focus effect
Multi-component · Fingerprint spectrum

1 Introduction

With the development of society as well as economy, automobiles have been greatly increasing. However, the environmental pollution resulting from the epilogue of automobiles is getting more and more severe. Meanwhile, as people's living conditions are improving, fitment becomes popular in general families and its released harmful gases to the living condition bring much attention. Therefore, atmospheric quality has

H. P. Urbach and Q. Yu (Eds.): ISSOIA 2017, SPPHY 209, pp. 18–25, 2018.
https://doi.org/10.1007/978-3-319-96707-3_3

become a principal concern throughout the world. Moreover, it has become a challenging issue as a result of the increase of energy consumption, urban population and automobiles. Accordingly, we are confronted with an urgent solution to the environmental pollution and the most important issue in ameliorating environmental pollution is to test the polluted air precisely, rapidly and effectively. Therefore, many countries are studying the technology of space-based atmospheric pollutants detection and developing the related equipments. Especially, the multi-component atmosphere detection has become the research focus, due to the scientific significance and application prospect. We need a new technology to detect the atmosphere.

2 Filament Laser

Ultrafast optics is the most advanced technology of instantaneous optics area. Ultrafast laser is related to the instantaneous phenomenon of the tiny time scale, which is the shorter than femtosecond. From the middle 1980's, as the important part of ultra-fast optics, ultra-fast laser has been developed quickly, which shows the application values in many fields. Ultra-fast laser has two significant features. One is short pulse duration, and the other is high peak power. Take the femtosecond laser for example. The pulse duration is the level of femtosecond, i.e. ~ 10–15 s. The peak power can reach the level of $\sim 10^{21}$ W/cm^2 [1] (Fig. 1).

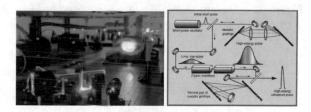

Fig. 1. The femtosecond laser technology

When the ultra-fast laser propagates in air, the air is ionized to plasma, because of high peak power. Due to the nonlinear physical effect, the plasma channel is generated. There are two main effects in this process, kerr self-focusing and plasma defocusing. When the two effects balance each other, the plasma channel can propagate a long distance [2]. Figure 2 shows the formation process of plasma channel when Kerr self-focusing balances Plasma defocusing. This complicated process can be described by self-guided model. When the femtosecond laser propagates in air, due to the third order electric polarization effect, the refractive index changes directly proportion to the incidence light intensity, as shown in (1):

$$n \approx n_0 + \frac{3Re(\chi^{(3)})}{2n_0}|E(\omega)|^2 \tag{1}$$

Where n_0 is the linear refractive index, $Re(\chi^{(3)})$ is the real part of third order nonlinear polarization rate. Take $\frac{3Re(\chi^{(3)})}{2n_0}$ as n_2, the Eq. (1) can be written as (2):

$$n \approx n_0 + n_2|E|^2 = n_0 + n_2I \tag{2}$$

Where n_2 is the nonlinear refractive index. For Gaussian beam, when the laser beam propagates in air, the intensities in optical axis and edge are different, which make the inhomogeneous distribution of medium refractive index to generate the propagating channel. This channel has high refractive index at center and low refractive index on edge, which makes the air like a lens to focus the laser beam, i.e. self-focusing process. The self-focusing process makes the laser intensity increase. When the intensity reach the threshold of air ionization, the air is ionized to generate the low density plasma [3]. The effect on refractive index can be described as (3):

$$\Delta n = -\frac{\omega_p^2}{2\omega^2} \tag{3}$$

Where $\omega_p = [4\pi e^2 n_e(I)/m_e]^{1/2}$ is the plasma frequency. Compared to the kerr self-focusing effect, the plasma's contribution to refractive index is negative, i.e. a negative lens, making the laser beam defocus. When the kerr self-focusing balances the plasma defocusing, the plasma channel generates in air. Figure 3 shows the filament laser in various scales. Figure 3(a) and (b) shows the filament laser in laboratory and in air, respectively. The filament laser imaged by telescope is showed in Fig. 3(c).

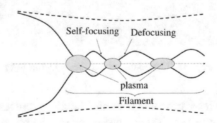

Fig. 2. The process of Kerr self-focusing balances Plasma defocusing

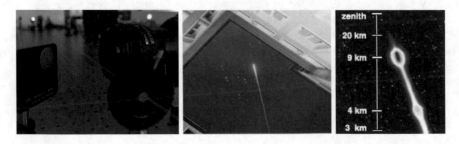

Fig. 3. Filament laser (a) in laboratory, (b) in air, (c) taken by telescope

3 The Concept

When the filament laser interacts with the air, the pollutants are dissociated into small fragments, and the molecules and atoms are ionized. Similar to the neon lights, the filament laser ionizes the molecules, and then the fluorescent spectrum can be generated according to the chemical substances. The fluorescent spectrum can be called "fingerprint spectrum", which can be applied in identifying the pollutants. In recent years, some experiments have been accomplished in laboratory, which shows that the detected distance can reach the level of kilometres [4]. Because of the characteristics of filament laser, it is possible to use the filament laser in space remote sensing, as shown in Fig. 4. In the concept, the filament laser can provide important information, which can be employed in the research of climate science and atmospheric chemistry [5].

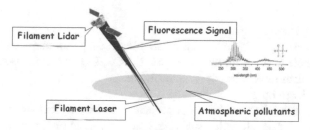

Fig. 4. Space-based filament laser system

Simulation by advanced concept team of European Space Agency (ESA), the filaments generate form the orbit of 400 km [6]. The numerical investigations provide new physical insight into nonlinear optical pulse propagation from a 400 km altitude towards the ground. The current results highlight the laser parameters required for an orbiting platform to remotely generate filament in the atmosphere. Figure 4 illustrates the beam radius evolution as a function of propagation distance, starting from orbit at $z0 = 400$ km. As shown in Fig. 5(a'–a''''), the beam considerably shrinks during the focusing stage until it reaches the nonlinear focus at ~ 10.015 km. Figure 5(a''') shows the formation of the filament once plasma defocusing and multi-photon absorption regularize the collapse and the subsequent oscillatory dynamics of the beam diameter. Figure 5(b–c) show the pulse intensity and plasma density at the beginning of filament (9.98 km $< z <$ 10.03 km), with or without nonlinear losses in the model. Focusing-defocusing cycles are clearly visible and reflect the competition between self-focusing and nonlinear plasma effects. Figure 5(a'''') shows that the filament length, defined as the distance between the nonlinear focus and the last intensity maximum at 10 TW/cm^2, is longer than 30 m.

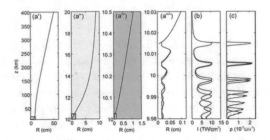

Fig. 5. Simulation results by ESA

Compared to other detection technologies as shown in Table 1, the filament lidar owns the better abilities to perform the detection obviously. The features are listed as follows:

- Variety of Measurement
  The detection targets include both of simple gaseous pollutants and complex pollutants. Simple gaseous pollutants are SO2, NO2, CO, etc., and complex pollutants are biochemical molecular, dust, aerosol, etc.
- Measuring large range
  The length of the filament laser can propagate as long as tens of kilometers, which puts the base for developing a spacecraft payload based on filament laser, as called as "Earth-Orbiting Filament Lidar".
- High precision detection
  The detection sensitivity is high, and the precision can reach 1 ppm.

Table 1. Compare between different detection systems

Detection types		Filament lidar	Traditional spectrometer	Differential absorption lidar
Lighting condition		Day and night	Day not night	Day and night
Matter to be detected	Unknown matter	√	√	×
	Absorption spectrum	√	√	Special line
	Fluorescence	√	√	Special line
	Matter without fluorescence	Ionization matter to detect	×	×
	Types and amounts	Multiple at the same time		One type usually

4 Study Works

4.1 Theory and Simulation

For the requirements of real application, the key point of filament laser is the length of propagation. In our study, some new methods have been researched to prolong the propagation length of filament laser, including the combination of an annular Gaussian beam and Gaussian beam. The annular Gaussian beam is the beam with zero center intensity. The two types of beams pass through an optical system, which is composed of an axicon and a plano-concave lens. By simulation, the difference of the nonlinear dynamics mechanism between the two types of laser beams is analyzed specifically [7]. The simulation results show that, under the same initial condition, the propagation length of a stable plasma channel induced by the annular Gaussian beam has a significant improvement, compared with the Gaussian beam, as shown in Fig. 6. In the meanwhile, the simulation numerically results demonstrate that the characteristics of the ring-Gaussian filament. Firstly, the energy deposition of filament laser pulse depends sensitively on the external geometry focusing and laser pulse parameters. Secondly, The length of filament laser can be adjusted obviously by setting the lens parameters, and the longest propagation occurs at the position near the maximum energy deposition. In addition, the research results find that, when the initial laser pulse intensity is the same, the length and uniformity of the plasma strings can be tuned by increasing the beam width, however the influence is not dramatic by increasing the beam radius. In other simulation, a method is proposed to calculate the fine structure of kinetic energy of filament laser, which bridges the two parts of researches of plasma channel usually studied independently of each other, i.e., the extending of the length of plasma filament and the prolonging of the lifetime of filament laser, as shown in Fig. 7 [8].

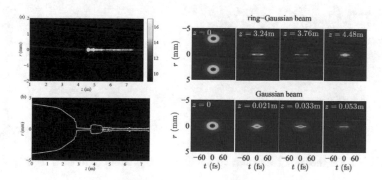

Fig. 6. The prolong of filament laser by ring-Gaussian beam

4.2 Experiment

In the lab, the experimental system has been setup, and the experiments include three main aspects. (1) Diagnosis of filament laser's lifetime, (2) Prolong of filament laser's length, (3) Detection ability of filament laser [9]. The lifetime of filament laser can be

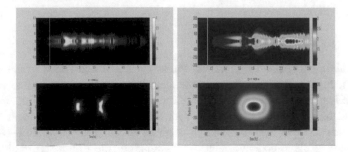

Fig. 7. Energy distribution of electrons in filament laser

measured by laser bump-detect interferometry, as shown in Fig. 8(a). The laser system we used in the experiment is a Ti:sapphire chirped pulse amplification laser system. Its central wavelength is around 800 nm. The pulse duration measured by an autocorrelator is about 50 fs. The femtosecond laser pulse is divided into two pulses by BS1. One pulse is bump beam, which generate the filament laser focused by L1. The other beam is detecting beam, reflected by M1 and M2, which detect the spectrum of filament laser. The spectrum from the air stimulated by filament laser is collected by L2 and L3, and analyzed by spectrometry to calculate the lifetime. Figure 8(b) shows the experimental setup of prolong of filament laser's length. Femtosecond laser pulse is divided into pulses with the same energy by BS1. Through delay line (M3, M4, M5 and M6), there is a time delay between two pulses. One pulse generates the filament laser, and other pulse propagates along the filament laser after time delay. Adjusting the time delay, the lifetime of filament laser is changed to prolong the length.

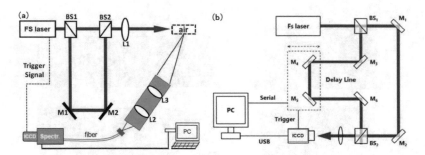

Fig. 8. (a) Diagnosis of filament laser's lifetime, (b) prolong of filament laser's length

On the base of above studies, the detecting experiment will be performed. The experimental setup is shown as Fig. 9. The gas to be detected is in the chamber. The filament laser propagates into the chamber through the window and interacts with the gas. The spectrum information is collected by the spectrometry. According to the spectral lines, the substances can be identified, such as, Freon, metal sample, protein powder, aerosol, etc. The analysis of the pollutant concentration can be performed through direct measurements of the characteristic spectral lines.

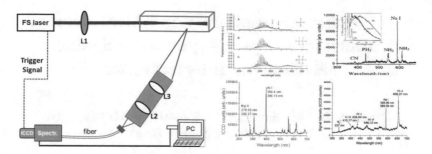

Fig. 9. Detecting experiment setup

5 Conclusion

In this paper, we prove the basis for a new technology to detecting the multi-components in atmosphere. Multi-component atmosphere detection technology based on filament laser shows the features of LIDAR, which can carry out the real-time vertical dimensional detection in large scale continuously. The technology can be used in daily environmental detection, early warning for atmospheric pollution, and sudden pollution survey, etc.

References

1. Kasparian, J., Rodriguez, M., Mejean, G., Yu, J., Salmon, E., Wille, H., Bourayou, R., Frey, S., André, Y.-B., Mysyrowicz, A., Sauerbrey, R., Wolf, J.-P., Woste, L.: White light filaments for atmospheric analysis. Science **301**, 61–64 (2003)
2. Braun, A., Korn, G., Liu, X., Du, D., Squier, J., Mourou, G.: Self-channeling of high-peak-power femtosecond laser pulses in air. Opt. Lett. **20**, 73–75 (1995)
3. Couairon, A., Mysyrowicz, A.: Femtosecond filamentation in transparent media. Phys. Rep. **441**, 47–189 (2007)
4. Bourayou, R., Mejean, G., Kasparian, J., Rodriguez, M., Salmon, E., Yu, J., Lehmann, H., Stecklum, B., Laux, U., Eisloffel, J., Scholz, A., Hatzes, A.P., Sauerbrey, R., Woste, L., Wolf, J.-P.: White-light filaments for multiparameter analysis of cloud microphysics. J. Opt. Soc. Am. B **22**, 369–377 (2005)
5. Rosenthal, E.W., Jhajj, N., Wahlstrand, J.K., Milchberg, H.M.: Collection of remote optical signals by airwaveguides. Optica **1**, 5–9 (2014)
6. Dicaire, I., Jukna, V., Praz, C., Milian, C., Summerer, L., Couairon, A.: Spaceborne laser filamentation for atmospheric remote sensing. Laser Photonics Rev. **10**(3), 481–493 (2016)
7. Scheller, M., Mills, M.S., Miri, M.-A., Cheng, W., Moloney, J.V., Kolesik, M., Polynkin, P., Christodoulides, D.N.: Externally refuelled optical filaments. Nat. Photonics **8**, 297 (2014)
8. Feng, Z.F., Li, W., Yu, C.X., Liu, X., Liu, J., Fu, L.B.: Extended laser filamentation in air generated by femtosecond annular Gaussian beams. Phys. Rev. A **91**, 033839 (2015)
9. Liu, X., Xin, L., Zhang, Z., Liu, X.-L., Ma, J.-L., Zhang, J.: Triggering of high voltage discharge by femtosecond laser filaments on different wavelengths. Opt. Commun. **284**, 5372–5375 (2011)

The Upper Atmosphere Wind Measurement Based on Doppler Asymmetric Spatial Heterodyne Interferometer

Hailiang Shi[1,2(✉)], Jing Shen[1,2,3], Zhiwei Li[1,2], Hailiang Luo[1],
Wei Xiong[1,2], and Xuejing Fang[1,2,3]

[1] Anhui Institute of Optics and Fine Mechanics,
Hefei Institutes of Physical Science, Chinese Academy of Sciences,
Hefei 230031, Anhui, China
hlshi@aiofm.ac.cn
[2] University of Science and Technology of China, Hefei 230026, Anhui, China
[3] Key Laboratory of Optical Calibration and Characterization of Chinese
Academy of Sciences, Hefei 230031, Anhui, China

Abstract. The wind detection in the upper atmosphere plays an important role in understanding the dynamics and actinochemistry, building the atmospheric dynamic model, offering the long term weather forecast and guaranteeing the development of aeronautics and astronautics. Doppler Asymmetric Spatial Heterodyne (DASH) technique is proposed to detect the wind speed in the upper atmosphere, depending on its wide field of view, large etendue, high spectrum resolution, static measurement and multiline capability. In this article the relationship between DASH optical path difference and the phase sensitivity and visibility is discussed. Then the principle prototype with large offset to detect Doppler shift of O_2 A-band airglow is described. The interferometric system is simulated and designed. The system noise and temperature change of the environment is simulated and analyzed quantitatively. After that the detailed parameters of the interferometer system is designed according to the requirement. Based on the instrument an experimental system is established in the laboratory. The wind simulation detection is conducted to validate the performance of the instrument. From a group of measurements under different wind speed, a wind precision of 3 m/s is obtained. The result shows that DASH interferometer is suited for passive wind detection.

Keywords: Upper atmosphere · Spatial heterodyne spectroscopy
Wind detection · Phase shift

1 Introduction

In the upper atmosphere, wind has significant influence on the transmission of energy, momentum and atmospheric composition. Besides, wind is an important environmental parameter that affects the spacecraft's orbit and safety [1–3]. The study of wind can lead to a better understanding of the upper atmosphere. Three techniques are used to the passive wind detection. The earliest one is using Fabry-Perot interferometer. Wind

© Springer Nature Switzerland AG 2018
H. P. Urbach and Q. Yu (Eds.): ISSOIA 2017, SPPHY 209, pp. 26–36, 2018.
https://doi.org/10.1007/978-3-319-96707-3_4

speed can be retrieved by the shift of fringe. It has high sensitivity but low etendue. What's more, the high flatness of etalon is demanded. As with the development of field-widened technique, Michelson interferometer is used in the wind measurement. Four points' intensity in the interferogram is used to calculated the phase and speed [4]. Although the Michelson interferometer has high etendue, only single emission line can be detected. Recent years, Doppler Asymmetric Spatial Heterodyne (DASH) is proposed to measure the wind speed. It has no moving parts, high etendue and can observe multiple emission lines simultaneously [5–7].

As is known to all, DASH has been a kind of mature technique since it was proposed in 2006 [5]. DASH is used on the thermospheric wind measurements firstly since the concept experiment was successfully validated in the laboratory [6]. The ground-based instrument called REDDI and the space flight prototype called Atmospheric Redline inteRferometer for dOppler Winds (ARROW) use the Koester prism as beamsplitter to form the common path interferometer. Although the quasi-common path configuration in REDDI is insensitive to translations of the interferometer elements and can easily be assembled as a monolith, the optical path difference (OPD) to retrieve the wind speed is limited. Therefore, it has low phase sensitivity. The wind detection ability of REDDI is validated together with a mature Fabry-Perot interferometer at the same time [7]. The result shows good overall agreement but large uncertainty.

Under the circumstances, the Large Offset DASH Interferometer (LODI) with double arms is designed. The OPD produced by one arm's offset is 12.29 cm. OPD is obtained by adding thickness of spacer in one arm. The prototype is used to detect the Doppler shift of the O_2 A-band airglow with compact structure and high phase sensitivity.

This paper introduces the basic principle of DASH technique firstly and then the choice of optimal path difference (L_{opt}). Next, the simulation of the system including the noise and temperature change are conducted. After that a glued interferometer is designed and the optical system for the experimental detection is built. Through the sampled interferogram, the phase shift caused by the Doppler shift of wavelength is calculated and the wind speed is retrieved.

2 Theoretics

2.1 Interferometric Function

The schematic is shown in Fig. 1. Light enters the beamspilliter after the collimator. The parallel light is divided into two beams. The transmission beam and reflected beam travel to the surfaces of two gratings after the field-widened prisms. Diffraction by the gratings, the beams come back to beamspilliter and form the Fizeau fringe pattern. Using an imaging lens the fringe is sampled on the CCD detector. The difference between DASH and SHS technique is the OPD introduced by one arm's offset of Δd. The interferometric function is written as [6]:

$$I(x) = \int\limits_{0}^{\infty} B(\sigma) \left[1 + \cos\left\{ 2\pi[4(\sigma - \sigma_L)\tan\theta_L]\left(x + \frac{\Delta d}{2\tan\theta_L} \right) \right\} \right] d\sigma \qquad (1)$$

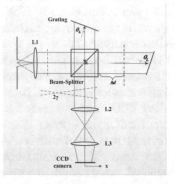

Fig. 1. Schematic of a DASH interferometer

Where x is the location on the detector array shown in Fig. 1 (x = 0 is the center of the array), σ is the wavenumber of the target emission line, $B(\sigma)$ is the spectral density of the incident radiation, σ_L is the Littrow wavenumber determined by σ_L, which is the Littrow angle of the gratings.

2.2 Choice of L_{opt}

Based on the optical Doppler theory, the wavenumber received by the detector is:

$$\sigma = \sigma_0 \left(1 + \frac{v}{c} \right) \qquad (2)$$

Where σ_0 is the wavenumber under zero wind speed, v is the wind speed in the atmosphere and c is the velocity of light.

From Eqs. 1 and 2, phase shift caused by the wind speed v is:

$$\delta\phi = 2\pi \cdot \frac{v}{c}\sigma_0 \cdot (4x\tan\theta_L + 2\Delta d) \qquad (3)$$

As we can see from Eq. 3, in the middle of the detector array where $x = 0$, phase shift changes linearly with the wind speed v. In order to obtain large phase shift $\delta\phi$, the optimal path difference $L_{opt} = 2\Delta d$ needs to be large enough. While for a purely temperature broadened line with Gaussian Line shape, $B(\sigma)$ is:

$$B(\sigma) = C\exp(-2\pi^2\sigma_D^2 L^2) \qquad (4)$$

Where C is a constant, σ_D is the line width ($\sigma_D = \sigma_0 \sqrt{\frac{kT}{mc^2}}$) and L is OPD.

Function 4 indicates that the increasing of L_{opt} leads to the decrease of envelope, which finally influences the quality of interferogram. Therefore, it is important to choose an appropriate offset L_{opt} to make a balance between phase shift and the visibility of interferogram.

In the upper atmosphere, temperature varies from 150 K to 1500 K [3]. Three typical temperatures are used to calculate the envelope. Besides, Michelson Interferometer for Global High-resolution Thermospheric Imaging (MIGHTI) based on DASH technique has proposed the highest wind precision of approximately 2 m/s [8]. Under

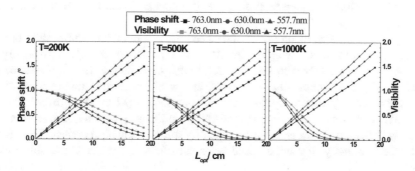

Fig. 2. Curves of phase and visibility changed with OPD under three different wavelengths. The left panel is calculated under the temperature of 200 K; the middle panel is calculated under the temperature of 500 K; the right panel is calculated under the temperature of 1000 K.

this wind precision, the phase and envelope curve with L_{opt} is calculated. The relationship of L_{opt}, phase shift and envelope of the interferogram can be shown in Fig. 2.

It is noted that the airglow in visible region are the main emission lines, as shown in Fig. 2. From the left panel, when the temperature in the upper atmosphere is 200 K, the phase shift linearly increases with L_{opt}, while the envelope exponentially decreases with L_{opt}. For wavelength of 763 nm, the intersection point of the two curves is at $L_{opt} = 9$ cm. The value of the phase shift and the envelope are both insured. On the other hand, as the increasing of wavelength, the optimal L_{opt} shifts on the large value. From three panels in Fig. 2, the optimal L_{opt} decreases when the temperature changes from 200 K to 1000 K.

In practice, the L_{opt} is chosen when $\exp(-2\pi^2\sigma_D^2 L^2) = 0.6065$, whose relationship shows in Eq. 5:

$$L_{opt} = \frac{1}{2\pi\sigma_D} \tag{5}$$

3 Simulation

3.1 Signal to Noise Ratio

By analysing the phase change of the fringes sampled by the detector, it is possible to reverse the wind speed. When the photoelectric conversion is done by CCD detector, the noise is superimposed on the interferogram. Since the radiated brightness of the target is small and the Doppler shift signal is only 10^{-8} to 10^{-6} wavelength, the noise simulation is particularly important for estimating the Doppler wind speed sensitivity of the instrument. In this paper, the Signal-to-Noise Ratio (SNR) is used to represent the detector noise. By simulating the interferogram with noise, the relationship between the SNR and the phase frequency shift curve is obtained according to the calculation formula [6].

Assuming that the theoretical wind speed is 100 m/s, the phase shift curve is obtained when the SNR changes from 10 to 400, as shown in Fig. 3. In theory, the phase shift varies linearly with the increase of OPD according to Eq. 3. As can be seen from Fig. 3(a), the nonlinearity of the curve at low SNR is large (the distortion at both ends of the curve is caused by the window convolution [8] resulting in a large wind speed inversion error. As the increase of SNR, the linearity of the phase shift curve is gradually increased. In order to further quantify the relationship between SNR and the wind speed accuracy, the wind speed error calculated from the phase shit curve is shown in Fig. 4.

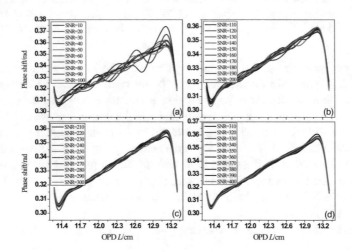

Fig. 3. The relationship between phase shift curve and SNR.

It can be seen from Fig. 4, when SNR is low the absolute value of the wind speed error fluctuates greatly. As the SNR increases, the absolute value of the wind speed error decreases. It is indicated that in the practical detection the system noise is required to be as small as possible and the signal to be as large as possible to reduce the

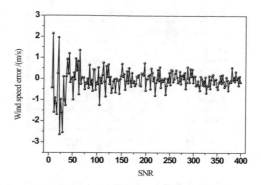

Fig. 4. The relationship between wind speed error and SNR.

inversion error. The SNR can be increased by cooling the detector, increasing the integration time and increasing the luminous flux.

3.2   Temperature Change

On the other hand, changes in ambient temperature can cause thermal drift of the instrument. The thermal effect of the DASH interferometer is not only due to the thermal change in the refractive index of the prism, but also the thermal expansion of the gratings in double arms. The change in the refractive index of the prism leads to the angle change of the outgoing light, resulting in an error in the Littrow angle of the gratings. The thermal expansion of the grating results in a change of the groove density. As can be seen from Eq. 6, changes in the Littrow angle and the groove density can lead to the error of Littrow wavenumber σ_L, thus affecting the wind speed inversion.

$$2\sigma_L \sin \theta_L = \frac{m}{d} \tag{6}$$

Where d is the period of grating line, m is the diffraction grade of the gratings and $m = 1$.

According to Eq. 3, the wind speed of 100 m/s can lead to the phase shift of 0.3 rad. From Fig. 5, it is indicated that the temperature change can leads to a large phase drift error that can't be ignored. Therefore, the wind speed retrieving is affected. When the instrument is designed, the glass material with smaller thermal coefficient is selected and the temperature controller are taken to minimize the phase drift error. In this test, the thermal drift was tracked and corrected by continuous sampling.

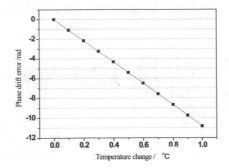

Fig. 5. Relationship between change of absolute phase at L_{opt} and temperature change.

4 Experiment

4.1 Instrument Configuration

Large Offset DASH Interferometer (LODI) based on DASH technique is designed to detect the Doppler shift of airglow in O_2 A-band and thus the wind speed in 80 to 105 km altitude can be calculated. It is a double-arm interferometer with large offset in one side. Figure 6 is the schematics of LODI. The interferometer consists of a cube beam splitter, spacers, field-widened prisms and gratings. Light beam is divided into two by the beam splitter and gets through the prisms. Diffracted by gratings the two beam go back to the beam splitter and then form the fringe pattern. Considering the practical situation simultaneously, L_{opt} of LODI is 12.29 cm. Using Eq. 3, the phase sensitivity is 3.4 mrad at 1 m/s. The main parameters are shown in Table 1.

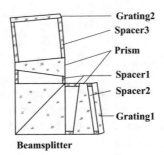

Fig. 6. Curves of phase and visibility changed with OPD under three different wavelengths.

An experimental set up is built to simulate the wind detection in Fig. 7. The experiment is mainly composed by the pre-shift device and interferometer system. Light from Potassium hollow cathode lamp (HCL) is collimated by lens. The parallel beam travels through a beam splitter and then reaches the Doppler wheel. Coating with a retro-reflecting tape on the plate, light beam returns back to the beam splitter along

Table 1. Design parameters of LODI

Beam splitter	Size (mm)	50 × 50
Spacer 1	Thickness (mm)	10
	Bottom angle (°)	77.5379
Prism (Silica)	Refraction	1.4539862@766.49 nm
	Thickness (mm)	11
	Bottom angle-α1 (°)	77.5379
	Bottom angle-α2 (°)	85.3692
	Apex angle (°)	17.0929
Spacer 2	Thickness (mm)	7
Spacer 3	Thickness (mm)	62
Grating	Groove density	600 g/mm
	Littrow angle (°)	13.1264
Scaling		0.621

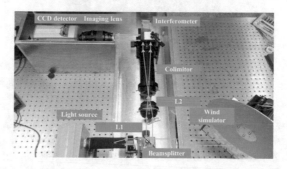

Fig. 7. Schematic of the LODI system

the same direction. When the plate rotates at a speed of n RPM, the beam will have a speed of:

$$v = 4\pi n r \cos(\alpha) \qquad (7)$$

Where r is the distance from light beam to the plate center, α is the angle between optical axis and plate. The wavenumber of the light source will shift as Eq. 8 is the distance from light beam to the plate center, α is the angle between optical axis and plate. The wavenumber of the light source will shift as Eq. 8

$$\sigma = \sigma_0 \cdot \frac{v}{c} \qquad (8)$$

Where σ_0 is the original wavenumber of light source. After that light beam enters the interferometric system and forms the fringe pattern. It is sampled on the CCD detector after scaling by the imaging lens.

4.2   Laboratory Measurement

The experiment is conducted in the laboratory with the environmental temperature of 25 °C. With a Potassium hollow cathode, two emission line 766.49 nm and 769.90 nm is in the pass band. The wavelength shift of light source is produced by Doppler wheel with a given speed and detected by LODI. The fringe number of the two emission lines are 73 and 192 respectively. Due to the dusts and scratches on every optical surfaces in the system, some dirty pixels are shown in the raw interferogram in Fig. 8. While it has small influence on the wind speed retrieval because this error can be eliminated on large scale when calculating the phase difference on two interferogram.

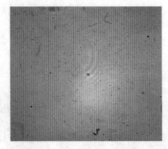

Fig. 8. Raw interferogram of double line with dark current corrected.

By controlling the rotating speed of motor, the wind speed can be simulated as Eq. 7. A series of interferogram frame is sampled with a total length of 2 N. The former part of interferogram frame is sampled under rotating speed of 0 RPM. The latter part of interferogram frame is sampled under rotating speed of M RPM (M ≠ 0). After calculating the fringe phase at L_{opt} for every single interferogram, a phase curve that changes with the frame can be got. The effective phase shift is easily to be derived from it.

5   Result

Based on the method, a group of interferogram was sampled to correct the phase drift and achieve the wind speed with high precision. In the laboratory, the Potassium hollow cathode lamp was started at 10 mA and the CCD detector was cooled to −40 °C. The integral time was 0.5 s with a total sampling length of 120. To keep a relative stable environment, the air-condition was closed when the room temperature was 25 °C (outside temperature was 35 °C). First, 50 frames of interferogram at 0 m/s were sampled continuously. Then the Doppler wheel was started to a stable rotating speed of 3600 RPM and 50 frames interferogram at 60.37 m/s were sampled again. It is noted that 10 s was cost to adjust the motor rotating speed during which 20 frames of invalid interferograms were also sampled. With the Fourier Transformation (FT) and Inverse Fourier Transformation (FT^{-1}), the absolute is achieved.

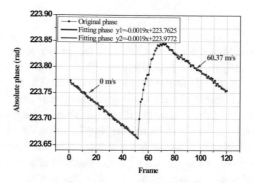

Fig. 9. Time series of the average phase at L_{opt} for a single interferogram. Black line is the phase curve at 60.37 m/s; Red line is the fitting curve of the formal 50 frames; Blue line is the fitting curve of the latter 50 frames.

The phase curve is divided into three parts which is corresponding to the sampling process described in the last paragraph. Despite the adjusting process of the rotate speed from 51 to 70 frames, it is indicated that the absolute phase decreases linearly with the sampling. It is caused by the decreasing of the Littrow wavenumber. By fitting the absolute phase curve using least square method, the function is shown in the left corner of Fig. 9.

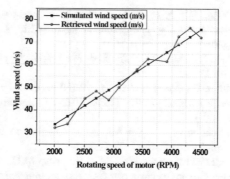

Fig. 10. The retrieved wind speed. Black curve is the simulated speed by the Doppler wheel. Red curve is the retrieved speed from the interferogram.

The wind speed was simulated in the laboratory from 33.75 m/s to 75.75 m/s using the Doppler wheel. The retrieved wind speed from the change of fringe phase (red curve) shows a good agreement with simulated wind. The mean error of the twelve groups of measurements is 2.97 m/s. It is shown that by correcting the drift using the linear fitting method, the wind can be retrieved with high precision (Fig. 10).

6 Conclusion

In this paper the relationship between OPD and the phase sensitivity and visibility is discussed first. The designing of L_{opt} is given based on the discussion. It known that system error such as noise and temperature change can lead to the wind retrieving error according to DASH technique. By simulating the noise it is indicated that the he wind speed error decreases with the increasing of noise. Thus through cooling the detector, increasing the integration time and increasing the luminous flux during the measurement the signal with high SNR can be got. On the other hand, the temperature change of the DASH interferometer that comes from the thermal effect of the optical elements contributes to the obviously phase drifts. To solve this problem, the thermal drift was tracked and corrected by continuous sampling. After the simulation above, the DASH interferometer was designed and the wind detection system was established in the laboratory. The measurement was made by simulating the wind speed (60.37 m/s) with a Doppler wheel (3600 RPM). The phase drift is fitted with a slope of −0.0019, as shown in Fig. 9. The phase shift is 0.2146 rad and the wind speed is retrieved with an error of 3.51 m/s. Several groups of measurements were conducted. The experimental measurement covered the wind speed from 33.75 m/s to 75.75 m/s and the results shows an average wind error of 2.97 m/s.

References

1. Wang, Y.J., Wang, Y.M., Wang, H.M.: Simulation of ground-based Fabry-Perot interferometer for the measurement of upper atmospheric winds. Chin. J. Geophys. **57**(6), 1732–1739. doi:10/6038/cig20140605
2. Abreu, V.J., Hays, P.B., Skinner, W.R.: The high resolution Doppler imager. Opt. Photon. News **2**(10), 28–30 (1991)
3. Shepherd, G.G., Thuillier, G., Gault, W.A., et al.: WINDII, the wind imaging interferometer on the upper atmosphere research satellite (1984–2012). J. Geophys. Res. Atmos. **98**(D6), 10725–10750 (1993)
4. Shepherd, G.G.: Application of Doppler Michelson imaging to upper atmospheric wind measurement. Appl. Opt. **35**(16), 2764–2773 (1996)
5. Englert, C.R., Harlander, J.M., Babcock, D.D., Stevens, M.H., Siskind, D.E.: Doppler Asymmetric Spatial Heterodyne Spectroscopy (DASH): an innovative concept for measuring winds in planetary atmospheres. SPIE Optics+ Photonics. International Society for Optics and Photonics, pp. 63030T–63030T-8 (2006)
6. Englert, C.R., Babcock, D.D., Harlander, J.M.: Doppler asymmetric spatial heterodyne spectroscopy (DASH): concept and experimental demonstration. Appl. Opt. **46**(29), 7297–7307 (2007)
7. Englert, C.R., Harlander, J.M., Emmert, J.T., Babcock, D.D., Roesler, F.L.: Initial ground-based thermospheric wind measurements using Doppler asymmetric spatial heterodyne spectroscopy (DASH). Opt. Express **18**(26), 27416–27430 (2010)
8. Shen, J., Xiong, W., Shi, H.L., Luo, H.Y., Li, Z.W., Hu, G.X., Fang, X.J.: Data processing error analysis based on Doppler Asymmetric Spatial Heterodyne (DASH) measurement. Appl. Opt. **56**(12), 3531–3537 (2017)

Snapshot Diffractive Optic Image Spectrometer: An Innovative Design for Geostationary Hyperspectral Imaging

Yun Su[1(✉)], Yanli Liu[1], Jianchao Jiao[1], Qiang Chen[2], Jianxin Li[2], and Chengmiao Liu[2]

[1] Beijing Institute of Space Mechanics and Electricity,
Key Laboratory for Advanced Optical Remote Sensing Technology of Beijing,
Key Laboratory for Optical Remote Sensor Technology of CAST, Beijing, China
suedul@163.com
[2] Nanjing University of Science and Technology, Nanjing, China

Abstract. Geostationary Orbit (GEO) staring imaging technology is an important field of aerospace remote sensing technology research. In this Letter, an efficient method and system for hyperspectral imaging of geosynchronous orbit is realized by fusing diffractive optic and light field imaging technology. Large scale axial scanning of diffractive optic image spectrometer (DOIS) limits its application in GEO satellite. The emergence of the light field imaging technology provides a perfect solution for DOIS. Our system is a snapshot spectrometer that project the spectral and spatial information simultaneously onto a CCD detector. Here a spectrometer system that operates in the visible near infrared band is designed and the performance of the system is analyzed and evaluated. Experiments are shown to illustrate the performance improvement attained by the new model. Our analysis shows that the novel snapshot hyperspectral diffractive optic image spectrometer is a high resolution, compact, economical, rugged, programmable, hyperspectral imager that can be widely built for visible near infrared detection camera and high sensitivity astronomy satellite in GEO orbit and other fields.

Keywords: Hyperspectral imaging · Diffractive optic element
GEO orbit · Light field camera

1 Introduction

Hyperspectral imaging, collects and processes information from across the electromagnetic spectrum. The goal of hyperspectral imaging is to obtain the spectrum for each pixel in the image of a scene, with the purpose of finding objects, identifying materials, or detecting processes [1, 2]. There are three general branches of spectral imagers for remote sensing. Prism or grating dispersive spectrometer, filter-based spectrometer and fourier transform spectrometer. Hyperspectral remote sensing is used in a wide array of applications, including precision agriculture, land surveying, environmental monitoring and most earth science disciplines.

© Springer Nature Switzerland AG 2018
H. P. Urbach and Q. Yu (Eds.): ISSOIA 2017, SPPHY 209, pp. 37–45, 2018.
https://doi.org/10.1007/978-3-319-96707-3_5

Binary optics is a surface-relief optics technology based on VLSI fabrication techniques. It has been intensively developed over the past decade as a new technology. The technology allows the creation of new, unconventional optical elements and provides greater design freedom and new materials choices for conventional elements. Binary optical elements (BOE), also known as phase zone plate lenses, have been proposed for many applications [3]. The diffractive optic element (DOE) has abundant chromatic aberration, providing dispersion as well as imaging. This capability allows designers to create innovative components that can solve problems in imaging spectrometer. The diffractive optic image spectrometer (DOIS) is a new way of geosynchronous earth orbit (GEO) hyperspectral imaging based on the staring array way instead of pushbroom imaging for LEO satellite [4]. With the increase in the lens aperture and widen in the spectrum, the dispersion range is also growing. Large scale axial scanning increases the difficulty of system design and manufacturing of the GEO spectrometer. Taking into account the domestic lens processing conditions constraints and detector scanning mechanism adjustment ability, designing a single lens to achieve dispersion as well as imaging becomes unrealistic.

With the rapid development of computational photography, a field that aims to improve capabilities and overcome limitations of conventional photography by combing imaging systems design method and image reconstruction algorithm, the emergence of the light field imaging technology provides a perfect solution for the diffractive optic image spectrometer of GEO orbit. In a single snapshot, The light field camera that records information from all possible view points within the lens aperture enable digital image refocusing and 3D reconstruction. It show that the light field camera obtain a larger depth of field but maintain the ability to reconstruct detail at high resolution [5].

In this Letter, an efficient method and system design for hyperspectral diffractive optic image spectrometer of geosynchronous orbit is realized by optical power distribution and light field imaging technology. The novel snapshot hyperspectral diffractive optic image spectrometer is a high resolution, compact, economical, rugged, programmable, hyperspectral imager that can be built for visible or infrared applications. Here the spectrometer system is designed and the performance of the system is analyzed and evaluated.

2 Principles

A schematic of the conventional diffractive optic image spectrometer (DOIS) system is depicted in Fig. 1. The diffraction lens builds images for different wavelengths at different distances. A CCD camera, mounted parallel to the DOE, is scanned along the optical axis recording a series of spectral images in a process referred to as diffractive spectral sectioning [4]. Under this project the 3D spectra and spatial information cube can be captured and processed.

The diffractive optic element (DOE) has abundant chromatic aberration. The spectral focal length of DOE is described by the constant K_0:

$$\lambda f(\lambda) = \lambda_0 f_0 = K_0 \tag{1}$$

$$f(\lambda) = \frac{\lambda_0 f_0}{\lambda} \tag{2}$$

Where f_0 is the focal length at a design wavelength λ_0.

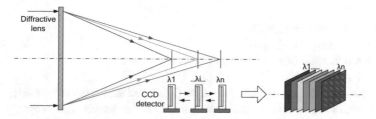

Fig. 1. Illustration of diffractive optic imaging spectrometer

The spectral resolution of the DOIS can be thought of as the ratio of the amount of in-band radiation to the amount of out-band radiation falling on a pixel. The intensity along the optical axis represents the expected amount of blurring applied to adjacent spectral slices. It is the limiting factor that determines the spectral resolution and bandwidth.

$$I(u, v) = h(u, 0) = \left(\frac{\sin \frac{u}{4}}{\frac{u}{4}}\right)^2 \tag{3}$$

The first zero in intensity along the axis is at

$$\Delta = 8\lambda (f/\#)^2 \tag{4}$$

According to (2),

$$df = \frac{\lambda_0 f_0}{\lambda^2} d\lambda \tag{5}$$

$$\Delta = \frac{\lambda_0 f_0}{\lambda^2} \Delta\lambda \tag{6}$$

The spectral resolution of the DOIS is

$$\Delta\lambda = \frac{\lambda^2}{\lambda_0 f_0} \Delta = \frac{\lambda^2}{\lambda_0 f_0} \cdot 8\lambda (f/\#)^2 = \frac{8\lambda_0 \lambda}{f_0} (f/\#)^2 \tag{7}$$

Where: f/# is the F number of the system.

Diffraction efficiency is the performance of DOE. It's a measure of how much optical power is diffracted into a designated direction compared to the power incident onto the diffractive element. The diffraction efficiency of quantitative structure DOE can be calculated from equation below. The diffraction efficiency will above 90% when N is 6 [4].

$$\eta_m^N = \frac{1}{N^2} \cdot \frac{\sin^2\left[\pi\left(1 - \frac{\lambda_0}{\lambda}\right)\right]}{\sin^2\left[\frac{\pi}{N}\left(1 - \frac{\lambda_0}{\lambda}\right)\right]} \cdot \mathrm{sinc}^2\left(\frac{m}{N}\right) \tag{8}$$

Where:
λ_0 is the design wavelength
λ is the spectral wavelength
m is the diffracted order of interest, typically m = 1
N is the number of levels in the DOE

The light field camera system shown in Fig. 2 includes a microlens array with the same f-number than the lens placed at lens focus. It's a camera with extended depth of field and records information from all possible view points within the lens aperture. It outputs light fields of the scene at a reduced spatial resolution of the detector in exchange of the angular resolution.4D spatial and angular information can be obtained after digital refocus process [6].

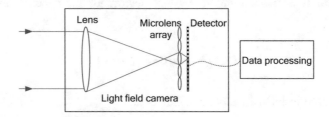

Fig. 2. Outline of the light field camera system

Classical radiometry shows that the irradiance from the aperture of a lens onto a point on the film is equal to the following weighted integral of the radiance coming through the lens.

$$E_F(x, y) = \frac{1}{F^2} \iint L_F(x, y, u, v) \cos^4 \theta \, du \, dv \tag{9}$$

where F is the separation between the exit pupil of the lens and the film, $E_F(x, y)$ is the irradiance on the film at position (x, y), L_F is the light field parameterized by the planes at separation F, and θ is the angle between ray (x, y, u, v) and the film plane normal. The $\cos^4\theta$ term is a well-known falloff factor sometimes referred to as optical vignetting. It represents the reduced effect of rays striking the film from oblique directions (Fig. 3).

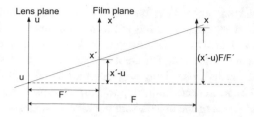

Fig. 3. Transforming ray-space coordinates

The diagram above is a geometric construction that illustrates how a ray parameterized by the x' and u planes for $L_{F'}$ may be re-parameterized by its intersection with planes x and u for L_F. By similar triangles, the illustrated ray that intersects the lens at u and the film plane at x', also intersects the x plane at u + (x' − u)F/F'. Although the diagram only shows the 2D case involving x and u, the y and v dimensions share an identical relationship. As a result, if we define $\alpha = F'/F$ as the relative depth of the film plane [6],

$$
\begin{aligned}
L_F(x', y', u, v) &= L_F\left(u + \frac{x' - u}{\alpha}, v + \frac{y' - v}{\alpha}, u, v\right) \\
&= L_F\left(u\left(1 - \frac{1}{\alpha}\right) + \frac{x'}{\alpha}, v\left(1 - \frac{1}{\alpha}\right) + \frac{y'}{\alpha}, u, v\right)
\end{aligned} \tag{10}
$$

This equation formalizes the 4d shear of the canonical light field that results from focusing at different depths. Combining (9) and (10) leads to the final equation for the pixel value (x', y') in the photograph focused on a piece of film at depth F' = αF from the lens plane [6]:

$$
E_{(\alpha, F)}(x', y') = \frac{1}{\alpha^2 F^2} \iint L_F\left(u\left(1 - \frac{1}{\alpha}\right) + \frac{x'}{\alpha}, v\left(1 - \frac{1}{\alpha}\right) + \frac{y'}{\alpha}, u, v\right) du\, dv \tag{11}
$$

This equation formalizes the notion that imaging can be thought of as shearing the 4D light field, and then projecting down to 2D.

3   Our System Model

With the increase in the lens aperture and widen in the spectrum, the dispersion range is also growing. Large scale axial scanning increases the difficulty of system design and manufacturing of the GEO spectrometer. In order to avoid the shortcoming of large scale axial scanning, a novel snapshot hyperspectral diffractive optic image spectrometer combing a diffractive lens, a refraction/reflection lens and a light field camera system is presented.

The schematic of the novel snapshot diffractive optic image spectrometer system can be seen in Fig. 4. The problem of large dispersion range and manufacture technique for large aperture of DOE is solved by the refraction/reflection lens, which bear the most focal power. The DOE provides dispersion and the refraction/reflection lens provides focus imaging in this system. The 3D spectra and spatial hybrid information can be captured by the light field camera in a single shot. Different images, which are a superposition of one wavelength focus and the defocused images of the other wavelengths, can be obtained by digital refocusing. The spectral decoupling algorithm can be used in reverse to reconstruct the accurate 3D image cube of the object from the blurry images.

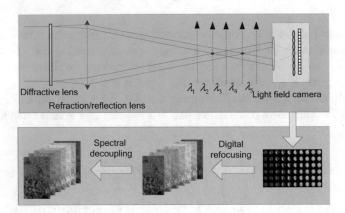

Fig. 4. Illustration of snapshot diffractive optic imaging spectrometer system

4 Optical Design

In order to verify experimentally the capability of the novel system for acting as a snapshot spectrometer, we use the optical design software ZEMAX to design the system. The integrated snapshot diffractive optic imaging spectrometer system was built by combining a diffractive lens, a coaxial three-mirror reflective optics system and a microlens array. The design parameters of the optical system are shown in Table 1. The optical design described in Fig. 5 and the microlens array design in Fig. 6 was built.

Table 1. Main parameters of the optical system

Items	Parameters
Wavelength	VNIR: 0.45–0.90 μm
Design wavelength	632.8 nm
Focal length	13 m
Entrance pupil diameter	1500 mm
F number	8.6
Field of view	0.45°

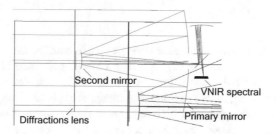

Fig. 5. Layout of diffractive optic imaging spectrometer

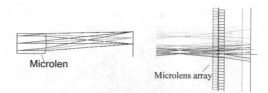

Fig. 6. Design of a microlens and microlens array

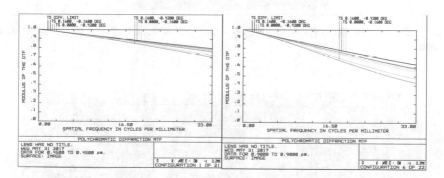

Fig. 7. Modulation Transfer Function curves of the visible near bands

The dispersion range of spectral coverage 450 nm–900 nm is 74.53 mm after the computational optimization. The depth of field (DOF) of the designed light field camera is 80.32 mm. All spatial and spectral information can be captured by the light field camera (Fig. 7).

According to the modulation transfer function curves of the visible near infrared band at respective fields and Nyquist frequency, MTF is above 0.4 when space frequency at 33 lp/mm. The minimum resolution requirements of the detector for optical system is satisfied.

5 Results

To verify the snapshot spectral imaging ability experimentally, we constructed a proof-of-concept prototype as shown in Fig. 8. The prototype consisted of a diffractive lens, refraction lens, light field camera lens, an array CCD detector with 1024×1024 pixels. A secondary imaging is designed to reimage the spectral image obtained by the microlens array onto the photosensitive surface of the detector. It is able to avoid the damage of the detector caused by the direct contact between microlens array and the detector and decrease the difficulty of adjustment and installation for the system as well. MATLAB routines were written to control and capture data on the CCD.

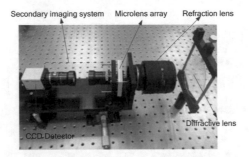

Fig. 8. Experimental prototype of the integrated snapshot diffractive optic imaging spectrometer system

Fig. 9. (a) The original scene (b) detector measurement of the scene (c) calibration image of microlens array

A test data cube with spatial resolution of 256×256 pixels and 28 spectral bands in the range of 450 nm to 720 nm was used. Figure 9 shows the original scene, the aliasing image captured by the detector and the calibration image of microlens array. Figure 10 depicts the aliasing spectral image refocused by the light field camera system using the refocus model described in Sect. 2. The spectral decoupled spatial content of the scene is showed also. Spectral resolution of the system is 10 nm through the spectral decoupling algorithm.

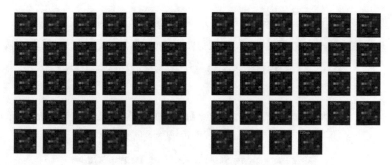

Fig. 10. (a) The digital refocused spatial content of the scene (b) The spectral decoupled spatial content of the scene in each of 28 spectral channels between 450 and 720 nm

6 Conclusions

The work presented here describes the design and analysis of a novel hyperspectral diffractive optic image spectrometer system for geosynchronous orbit. The diffractive optic element provides dispersion and the light field camera captures spatial and spectral information simultaneously compared to conventional camera's pushbroom imaging. The ability to simultaneously characterize a scene in 3 dimensions: 2 spatial and spectral makes us believe that this imaging methodology expands the space for spectrometer design and application. This system can be loaded at small speedy responsive space satellites or micro-satellite for hyperspectral reconnaissance, such as moving object real-time monitoring and ground military targets classification.

References

1. Chang, C.-I.: Hyperspectral Imaging: Techniques for Spectral Detection and Classification. Springer, Berlin (2003). ISBN 978-0-306-47483-5
2. Grahn, H., Geladi, P.: Techniques and Applications of Hyperspectral Image Analysis. Wiley, Hoboken (2007). ISBN 978-0-470-01087-7
3. Farn, M.W., Veldkamp, W.B.: Binary Optics, pp. 2–3. MIT/Lincoln Laboratory, Lexington (1995)
4. Lyons, D.M.: A Diffractive Optic Image Spectrometer (DOIS). Rome Laboratory Air Force Materiel Command Rome, New York (1997)
5. Bishop, T.E., Favaro, P.: The light field camera extended depth of field, aliasing, and superresolution. IEEE Trans. Pattern Anal. Mach. Intell. **34**(5), 972–986 (2012)
6. Ng, R.: Digital Light Field Photography. Stanford University, Stanford (2006)

Optimal Design of Dynamic Characteristic of Integrated Design Satellite Structure Based on Transmission Property Analysis

Zhenwei Feng[✉], Yongqiang Hu, Yufu Cui, Xinfeng Yang,
and Jiang Qin

Dong Fang Hong Satellite Co. Ltd, Beijing, China
18810235621@163.com

Abstract. With the development of remote sensing satellite technologies, and the pressing demand for high resolution and high agility, integrated design of platform and payload has been more pervasive, which could remarkably reduce costs, volume and mass of satellites. However, compared to traditional designs, remote sensors in the integrated design are much closer to vibration sources, implying that the micro-vibration environment of remote sensors becomes even worse. Although the micro-vibration is so small that its influence to structure is negligible, it could result in considerable reduction of image quality of high resolution remote sensors. Micro-vibration has become one of the critical factors for high resolution image quality especially in the integrated remote sensing satellites. However, the transmission property of micro-vibration in the satellite structure has not been well explained. In this article, the relationship between transmission property and modal parameter were studied to optimize the dynamic characteristic of integrated design satellites. First, the relationship between transmission property and modal parameter were derived and different suggestions were discussed for different conditions. Second, the system integrated FEM model was established to figure out the influence of global mode and local mode on transmission property and so the key link influencing the transmission property could be found. Finally, several suggestions were proposed on frequency assignment, configuration and layout and so on to optimize the dynamic characteristic and can provide reference to in subsequent integrated design satellites.

Keywords: Integrated design · Micro-vibration · Transmission property
Dynamic optimization

1 Introduction

With the development of satellite technologies, and the pressing need for high resolution, high agility and low costs, integrated design of satellite platform and payload has been a growing tendency. Compared to traditional designs, remote sensors in the integrated design are much closer to disturbance sources, implying that the micro-vibration environment of remote sensors becomes even worse. Micro-vibration has become a critical factor for image quality as to high-resolution satellites. Analyzing the

© Springer Nature Switzerland AG 2018
H. P. Urbach and Q. Yu (Eds.): ISSOIA 2017, SPPHY 209, pp. 46–54, 2018.
https://doi.org/10.1007/978-3-319-96707-3_6

whole propagation process from momentum wheels (MW), cryocoolers and other disturbance sources to sensitive payloads, there are three possible ways to suppress the vibration transmitted to payloads: cutting down the micro-vibration of disturbance sources, accelerating the attenuation along vibration paths, lowering the sensitivity of payload. However, considering current abilities of designing, manufacturing and testing, focusing on the propagation pathways is the most feasible way currently [1].

In the current research of micro-vibration, many tests and simulations have been carried out, attaining a mass of test and simulation data [2]. However, the transmission property of micro-vibration in the satellite structure has not been well explained, such as the theoretical study of relationship between transmission property and modal parameter, the optimization of frequency assignment among whole satellite, disturbance sources and sensitive payloads, the influence of local structure on transmission property [3, 4].

In this article, based on the micro-vibration analysis of the cryocooler of a certain remote sensing satellite, the relationship between transmission property and modal parameter will be studied to figure out the key link influencing the transmission of micro-vibration, combined with system integrated FEM model. So some suggestions can be offered on the frequency assignment of whole satellite, configuration and layout and so on to optimize to dynamic characteristic of satellite structure and can provide reference to subsequent design of satellite structure.

2 Transmission Property Analysis

Considering a 6n degrees of freedom satellites, its dynamic equation can be written as:

$$[M]\{\ddot{x}\} + [C]\{\dot{x}\} + [K]\{x\} = \{F(t)\} \tag{1}$$

Thereinto, [M], [C] and [K] are respectively mass matrix, damping matrix and stiffness matrix, which are all $6n \times 6n$ dimension. F(t) is the exciting force and $\{\ddot{x}\}$, $\{\dot{x}\}$, $\{x\}$ are respectively the acceleration, velocity and displacement. When the system is of free vibration, the dynamic equation can be written as:

$$[M]\{\ddot{x}\} + [K]\{x\} = \{0\} \tag{2}$$

As the amplitude of micro-vibration is particularly small, the solution of the equation can be defined as:

$$\{x\} = \{\Phi\}\sin(\omega t + \alpha) \tag{3}$$

Substituting (3) into (2), the equations to solve the natural frequency and modal shape can be written as:

$$([K] - \omega^2[M])\{\Phi\} = 0 \tag{4}$$

The nature frequency ω_i and modal shape $\{\Phi_i\}$ can be obtained by solving (4). Regularized modal shapes $\{\Phi_i\}$ can be assembled as modal shape matrix $[\Phi]$, which satisfies [5]:

$$[\Phi]^T[M][\Phi] = [\bar{M}] = [I] \tag{5}$$

$$[\Phi]^T[K][\Phi] = [\bar{K}] \tag{6}$$

Thereinto, $[\bar{M}], [\bar{K}]$ are respectively generalized mass matrix and generalized stiffness matrix, which are all diagonal matrix. To find out the influence of mode on transmission property, the dynamic Eq. (1) can be transformed by $\{x\} = [\Phi]\{\eta\}$:

$$[M][\Phi]\{\ddot{\eta}\} + [C][\Phi]\{\dot{\eta}\} + [K][\Phi]\{\eta\} = \{F(t)\} \tag{7}$$

Each side multiplies $[\Phi]^T$:

$$[\Phi]^T[M][\Phi]\{\ddot{\eta}\} + [\Phi]^T[C][\Phi]\{\dot{\eta}\} + [\Phi]^T[K][\Phi]\{\eta\} = [\Phi]^T\{F(t)\} \tag{8}$$

$$\{\ddot{\eta}\} + [\bar{C}]\{\dot{\eta}\} + [\bar{K}]\{\eta\} = [\Phi]^T\{F(t)\} \tag{9}$$

Then the coupled equation can be transformed into independent equation set:

$$\ddot{\eta}_i + c_i\dot{\eta}_i + k_i\eta_i = [\Phi_i]^T\{F(t)\} \tag{10}$$

Supposing there is only one excitation force $F_0 sin\ \omega t$ on p degree of freedom. The above equation can be written as:

$$\ddot{\eta}_i + c_i\dot{\eta}_i + k_i\eta_i = \Phi_{pi}F_0 sin\ \omega t \tag{11}$$

(11) is the standard single degree of freedom dynamic equation. The solution of this equation is:

$$\eta_i = \frac{\Phi_{pi}F_0}{\sqrt{m_i^2\left(\omega_i^2 - \omega^2\right)^2 + (c_i\omega)^2}} sin(\omega t - \varphi) \tag{12}$$

Thereinto: $\varphi = \arctan\dfrac{c_i\omega}{m_i\left(\omega_i^2 - \omega^2\right)}$. Then the response can be written as:

$$\{x\} = [\Phi]\{\eta\} = \sum_{i=1}^{6n} \eta_i\{\Phi_i\} = \sum_{i=1}^{6n} \frac{\Phi_{pi}F_0}{\sqrt{m_i^2\left(\omega_i^2 - \omega^2\right)^2 + (c_i\omega)^2}}\{\Phi_i\}sin(\omega t - \varphi) \tag{13}$$

So the response of q degree of freedom is:

$$\{x\}_q = ([\Phi]\{\eta\})_q = \sum_{i=1}^{6n} \eta_i \Phi_{qi} = \sum_{i=1}^{6n} \frac{\Phi_{pi} \Phi_{qi} F_0}{\sqrt{m_i^2 \left(\omega_i^2 - \omega^2\right)^2 + (c_i \omega)^2}} \sin(\omega t - \varphi) \quad (14)$$

It can be seen from (14):

(a) The transmission property of micro-vibration in the satellite structure is relate to the natural frequency and modal damping of the whole satellite and the modal shape of input and output degree of freedom.

(b) It is generally agreed that the modal damping of satellite structure is very small. So if the exciting frequency is not coupled with the modal frequency, the elastic force plays the leading role that the transmission property mainly determined by the difference between exciting frequency and modal frequency. So when we assign the frequency, the frequency of disturbance and the frequency of the whole satellite and sensitive payload need to separate as much as possible.

(c) When the exciting frequency is coupled with the modal frequency, the damping force plays the leading role. The additional damping structure should be considered to provide more inherent damping to improve the vibration resistance ability and stability of structure.

(d) The modal shape of input and output degree of freedom has great influence on transmission property. If $\Phi_{qi} = 0$, which means that the output node is precisely at the modal node, the contribution of this mode to the response of this degree of freedom is zero.

What needs to be explained is that the response under single sine excitation force has been discussed on the above and as to the general periodic excitation forces, it can be expanded into Fourier series, and the response of each component can be solved. It is generally agreed that the structure of satellites is linear. The whole response can be obtained by superposing the solution of each component, according to the superposition principle of linear system.

3 Analysis and Optimization of Influence of Modal Parameter on Transmission Property

In a certain high-resolution satellite, the cryocooler mounts on the infrared camera and the infrared camera and high-resolution camera are both mount on the payload panel. The propagation path of micro-vibration from the cryocooler to the high-resolution camera is: cryocooler on the infrared camera → mounting point of infrared camera → mounting point of high-resolution camera → the flange of main mirror of high-resolution camera → second mirror of high-resolution camera. The disturbance of the cryocooler is showed in Fig. 1, which peak in frequency domain is 54 Hz and its integer multiple.

As the optical system is not sensitive to the displacement of yaw axis (z direction), only the response of x and y direction is considered in the following analysis. Four critical points on the propagation path are selected to analyze the transmission property

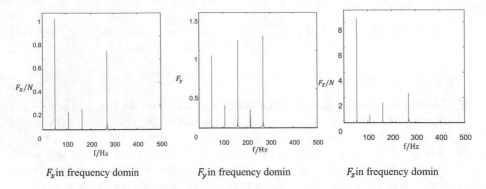

F_x in frequency domin F_y in frequency domin F_z in frequency domin

Fig. 1. The disturbance of the cryocooler in frequency domain

of micro-vibration. As shown in Fig. 2, there are obvious peak around 22 Hz, 24 Hz, 69 Hz at all four critical points. These three frequency are exactly the natural frequency of the satellite. The main exacting frequency 54 Hz, 108 Hz are not coupled with natural frequency of satellite and so the transmission ratio is small, which means that the frequency assignment is reasonable. In addition, the 162 Hz component is amplified at the second mirror of high-resolution camera. The structure of high-resolution may be the most probable cause and it will be discussed in the Sect. 3.2.

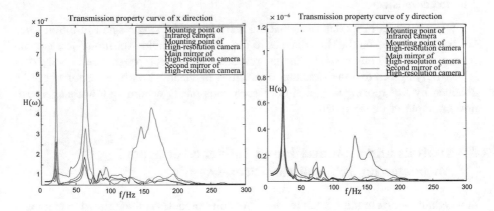

Fig. 2. Transmission property of four critical points along the propagation path

3.1   Influence of Global Mode on Transmission Property

Figure 3 Present the displacement of second mirror in frequency domain. As shown in, 54 Hz and 162 Hz component are the main disturbance frequency. So in this section, the 54 Hz will be studied to figure out the influence of global mode on transmission property, as shown in Fig. 4.

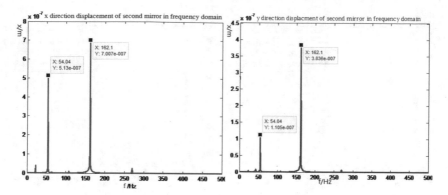

Fig. 3. The displacement of second mirror in frequency domain

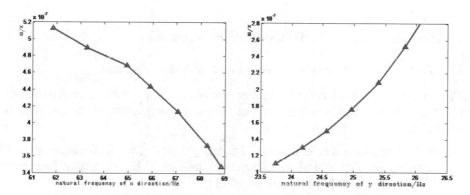

Fig. 4. The 54 Hz response of second mirror varies with the mode of whole satellite

As shown in Fig. 4, as the natural frequency of x direction goes up, the 54 response of second mirror displacement gets small. Getting Fig. 2 into consideration, as the natural frequency of x direction goes up, the transmission ratio goes down and so the response goes down. The variations of y direction is also caused by global mode.

3.2   Influence of Local Mode on Transmission Property

As shown in Fig. 2, the 162 Hz component is amplified at the second mirror of high-resolution camera, but it is not amplified from the cryocooler to the main mirror of high-resolution camera and it is verified in the micro-vibration test of whole satellite. In order to find out the reason of the amplification, further analysis have been done to the propagation path from main mirror to the second mirror, as shown in Fig. 5. When the micro-vibration transmits to the main mirror and the root of the barrel, the 162 Hz component is not amplified, but at the root of three rod and the second mirror, the 162 Hz component is amplified. What makes the amplification is that the 162 Hz component is coupled with the natural frequency of barrel.

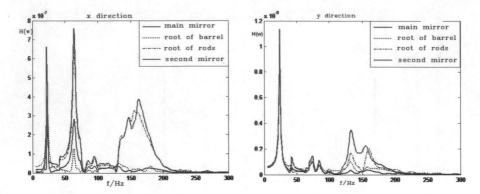

Fig. 5. Frequency response characteristic on the pathway from the primary mirror to the second mirror

4   Optimization of the Dynamic Characteristic

4.1   Optimization of Frequency Assignment of Whole Satellite

As shown in Fig. 2, the disturbance frequency is not coupled with the natural frequency of whole satellite, which means that the frequency assignment of this satellite is acceptable.

(1) In the range of x direction natural frequency ± 10 Hz, the transmission ratio is quite big. So if the disturbance frequency keeps on 54 Hz, the natural frequency of x direction should be greater than 64 Hz.

(2) In the range of y and z direction natural frequency ± 7 Hz, the transmission ratio is quite big. So if the natural frequency of whole satellite does not change, the frequency of disturbance should satisfy:

$$\begin{cases} 31\,\text{Hz} < x < 52\,\text{Hz} \\ 72\,\text{Hz} < x < 96\,\text{Hz} \end{cases} \tag{15}$$

As shown in (15), when the frequency of disturbance is in the range of 36–48 Hz, the micro-vibration environment of high-resolution camera could be better.

4.2   Optimization of Transmission Property of Critical Structure

As mentioned earlier, the amplification of 162 Hz component is caused by the coupling of natural frequency of barrel. So in order to analyze the influence of stiffness of barrel on the transmission property, the modulus of elasticity is changed, as shown in Figs. 6 and 7. After the change, the amplification of 162 Hz gets efficiently control, especially on x direction. The response of x direction and y direction both reduced more than 66%.

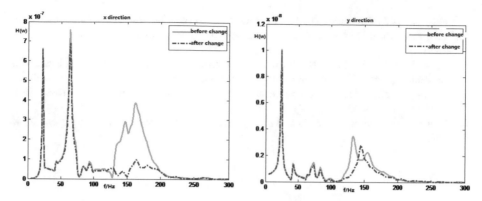

Fig. 6. Comparison of transmission property before and after the change of material of barrel

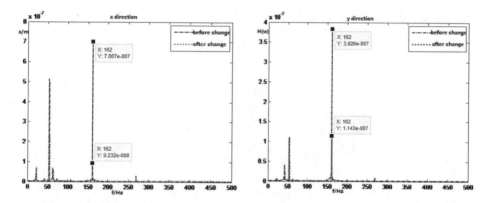

Fig. 7. Comparison of response of the second mirror before and after the change of material of barrel

5   Conclusion

In this article, the propagation process of attenuation and amplification along vibration pathways is analyzed based on the influence of mode on transmission property so that several suggestions are proposed on frequency assignment, configuration and layout and so on to optimize the dynamic characteristic and can provide reference to in subsequent integrated design satellites.

References

1. Feng, Z., Cui, Y., Yang, X., Qin, J.: Micro-vibration issues in integrated design of high resolution optical remote sensing satellites. In: ISSOIA, vol. 192, pp. 459–469. Springer Proceedings in Physics (2017)

2. Yang, X., Bai, Z., Yang, D., LI, Y.: Study on micro-vibration of satellite induced by momentum wheels and on-orbit simulation analysis. Equip. Environ. Eng. China **12**, 15–21 (2015)
3. Zhang, Z., Yang, L., Pang, S.: Analysis of micro-vibration environment of high-precision spacecraft. Spacecr. Environ. Eng. China **26**, 528–534 (2009)
4. Liu, C., Jing, X., Daley, S., Li, F.: Recent advances in micro-vibration isolation. Mech. Syst. Signal Process. **56**, 55–80 (2014)
5. Chengjun, W.: Vibration and Control in Engineering. Xi'an Jiaotong University Press, Xi'an (2008)

Impact Research of Drift of the Local Time of Descending Node on the Imagery of CCD Camera

Lingyan Gao$^{(\boxtimes)}$, Jinqiang Wang, Yong Liu, and Shaoyuan Cheng

Key Laboratory for Advanced Optical Remote Sensing Technology of Beijing,
Beijing Institute of Space Mechanics and Electricity, Beijing 100094, China
gaolingyan20080910@126.com

Abstract. The orbit plane's orientation of the satellite in sun synchronous orbit relative to the sun is settled, i.e., the local time of descending node (hereinafter named as LTDN) of the satellite is almost steady. The LTDN of earth observing remote sensing satellites such as IKONOS, SPOT, Worldview, ZY-3, VRSS-1 and GF-1 selected 10:30 am. However, the LTDN will shift on account of the influence of many aspects, such as satellite's orbit injection error. The LTDN shift of domestic remote sensing cameras (CBERS, ZY, GF-1, etc.) is controlled in ±30 min. The deviation of local time will have an impact on imaging quality. Based on the fact, a research and analysis approach of impact on the imaging radiometric quality of CCD camera resulted from the LTDN's drift is presented. By comparing the calculation results, the impact on remote sensing CCD camera's imaging can be obtained. The contrasts result shows that the imaging performance in observation period of a year and latitude observation scope restricted by solar altitude angle, and imaging SNR has been improved with the LTDN's shift towards positive values and has a reduction with the LTDN's shift towards negative values. The maximum variation in these performance parameters under the condition of the LTDN's drift in ±30 min can reach about 15 days, 2.2°, and 0.93 dB. Study and analysis result in this paper can give a quantitative reference to the CCD camera's imaging quality analysis. Meanwhile, it can also provide a data support to the optimal design orbit and the shift control strategy of LTDN in future remote sensing satellite system.

Keywords: Visible light CCD camera · The local time of descending node
Solar altitude angle · Earth observation · The radiance at the entrance pupil
SNR

1 Introduction

Optical remote sensing satellite imagery is acquired by receiving solar radiation reflected from the ground and target radiation in sun synchronous orbit [1]. The important character of sun synchronous orbit is that the sun illumination direction on orbit plane is almost steady in a year and anniversary repetitive. And the local time of descend node (hereinafter named as LTDN) of the satellite's flight across equator is not influenced. This character can meet the performance on earth observation of the

© Springer Nature Switzerland AG 2018
H. P. Urbach and Q. Yu (Eds.): ISSOIA 2017, SPPHY 209, pp. 55–64, 2018.
https://doi.org/10.1007/978-3-319-96707-3_7

satellite. Remote sensing satellite has the nearly same sun illumination when imaging the same latitude scene in the neighboring orbits. Then observation images data of the relevant areas can be obtained time-lapse, contrasted, analyzed and applied [2].

In respect that satellite's orbit injection error, gravitational force, atmospheric drag and sun gravitational perturbation have an impact on the satellite, the local time of descend node will drift [3]. And the deviation of local time will have an impact on satellite imaging quality. ZHANG Guoyun and his team analyzed the LTDN drift characteristics and effect caused by inclination deviation of nearly round sun synchronous orbit satellite, and sun illumination angle drift with different LTDN was given [4]. However, the in-depth study on impact of remote sensors' imaging quality due to the LTDN drift did not carried further.

This paper mainly researched on the sun synchronous orbit satellite push-broom TDICCD remote sensing panchromatic spectrum covering 500 nm to 800 nm imaging. By the process and mechanism study of the LTDN drift influencing on remote sensors' imaging quality, the imagery change that LTDN' drift brings has been simulated and calculated, analysis conclusion which are drew from the calculation results provides a reference to the remote sensor preliminary design, imaging quality demonstration and evaluation.

2   Survey of Research Process

The changes on remote sensing imagery caused by the LTDN drift, which includes intensity of illumination, the length of imaging time available in a year, imaging coverage latitude range and radiance of entrance pupil, can be simulated and computed gradually by satellite orbit simulation software and atmospheric radiation transmission software (6S: Second Simulation of the Satellite Signal in the Solar Spectrum). Then using CCD camera's imaging signal-to-ratio calculation model, the influence on imaging SNR can be calculated by MATLAB software. According to the results of imaging characteristics about the observation period in a year, imaging coverage latitude range and imaging SNR, this paper make an influence analysis and conclusion on remote sensing imaging quality caused by the LTDN drift (Fig. 1).

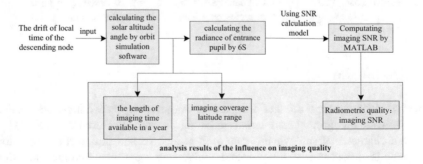

Fig. 1. Research and analysis process

3 Solar Altitude Angle Variation

The solar altitude angle (hereinafter named as SAA) is defined as the angle between the sunlight from the sun to the ground and local horizontal plane. The SAA ranges from 0° to 90°, and equals 90° when the sun is at the zenith. SAA varies with different local time and different latitude. Given the local latitude, year, month, day and local time, the SAA at that time can be computed [5, 6].

The value of SAA expressed as H_s equals the altitude above the horizon of the sun in the horizontal coordinate system, and it varies with the variation of local time and declination. δ is the declination, φ is the observation geographical latitude, t is the local time. Then the solar altitude angle [2] is

$$\sin H_s = \sin \varphi \cdot \sin \delta + \cos \varphi \cdot \cos \delta \cdot \cos t$$

In order to get better radiation flux and angle, the LTDN for optical remote sensing mission is generally around 9:00–11:00 am or 1:00–3:00 pm. The orbit simulation software was used to analyze the SAA variation resulting from the LTDN drift of ±30 min at different latitudes regions of the optical remote sensing satellite in sun synchronous orbit whose orbital altitude is 500 km, and the LTDN is 10:30 am. The analyzing results are shown in Figs. 2 and 3. The SAA variation in 4 typical solar terms of spring equinox, summer solstice, autumnal equinox and winter solstice at the same latitude is shown in the Table 1.

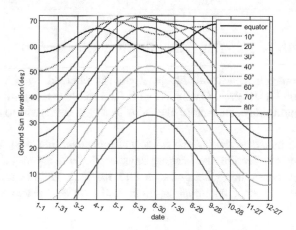

Fig. 2. The SAA at different latitudes in north hemispheres in a year resulting with LTDN of 10:30 am

The simulated data of Figs. 2, 3 and Table 1 shows that LTDN drift of ±30 min brings the SAA variation of imaging area. It is also seen that the SAA variation is different with different latitudes. The largest SAA variation appears in the summer solstice, and the SAA variation in the summer solstice at latitude 30° in north hemispheres is 6.3°.

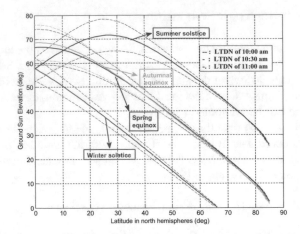

Fig. 3. The SAA variation at different latitudes with LTDN drift in ±30 min of 4 typical solar terms

Table 1. The SAA of different LTDN at latitude 30° in north hemispheres (units: °)

LTDN	Spring equinox	Summer solstice	Autumnal equinox	Winter solstice
10:00 am	50.2	65.0	50.8	30.7
10:30 am	54.5	71.3	54.8	33.5
11:00 am	57.8	77.3	57.7	35.4

4   The Length of Imaging Time Available in a Year

The length of imaging time available in a year is affected by the SAA variation resulting from the LTDN drift. Combined with the remote sensing mission request of the satellite, imaging days statistic of the SAA > 20° is important to remote sensing research.

Based on the analyzed data of the SAA variation resulting from the LTDN drift shown in the previous section, the imaging days available in a year of the SAA > 20° at different latitudes in north hemispheres and imaging latitude regions in spring equinox, summer solstice, autumnal equinox and winter solstice were calculated (Tables 2, 3).

Table 2. Statistics for SAA > 20° at different latitudes in north hemispheres (units: days)

Latitude	10:00 am	10:30 am	11:00 am
20°	All year	All year	All year
30°	All year	All year	All year
40°	All year	All year	All year
50°	280	293	301
60°	228	234	237

Table 3. Imaging latitude regions in 4 typical solar terms

LTDN	Spring equinox	Summer solstice	Autumnal equinox	Winter solstice
10:00 am	69.3°	>85°	68.6°	42.7°
10:30 am	69.9°	>85°	69.2°	44.9°
11:00 am	70.1°	>85°	69.3°	46.0°

Analysis results above shows that:

(1) in north hemispheres, the imaging period is through all year at the latitude 20°–40° region even with the LTDN drift, but has a reduction with the LTDN drift from 10:30 am to 10:00 am and has a promotion with the LTDN drift from 10:30 am to 11:00 am at the latitude 50°–60° region.

(2) Imaging coverage latitude region became broaden with the LTDN drift from 10:30 am to 11:00 am, and became narrow with the LTDN drift from 10:30 am to 10:00 am. the largest variation is 2.2° in winter solstice.

(3) In the ±30 min of LTDN drift, imaging coverage latitude region in summer solstice is the largest which covers over latitude 85°, and in winter solstice is the narrowest which only covers latitude 42.7°–46°.

5 Imaging SNR Variation of Camera

5.1 The Radiance at the Entrance Pupil of Remote Sensor

Spectral radiant sterance, I, is the basic quantity from which all other radiometric quantities can be derived. It is the amount of radian flux, (watts), per unit wavelength (micrometers) radiated into a cone of incremental solid angle (steradian) from a source whose area is measured in meters [7].

$$L = \frac{dI}{dA \cdot \cos\theta} = \frac{d^2\phi(\lambda)}{dA \cdot d\Omega \cdot \cos\theta}$$

Where dI is the radiant intensity, Ω is the solid angle, A is source area, $\Phi(\lambda)$ is radian flux. The units of L is $W/(m^2.Sr.\mu m)$.

The radiance at the entrance pupil of remote sensor with the SAA of different LTDN is calculated and analyzed by the atmospheric radiation transmission software 6S [8]. The calculation condition is:

(1) orbital altitude is 500 km
(2) visibility is 23 km
(3) atmospheric model is mid-latitude winter/summer
(4) aerosols type is continental aerosols model
(5) ground reflectance ranges from 0.05 to 0.65
(6) spectrum band is 500 nm–800 nm

The radiance at the entrance pupil in 4 typical solar terms of a year at the same latitude region is calculated. The radiance has the similar linear varying tendency with ground reflectance in different solar terms, therefore, only the data curve of spring equinox is shown in Fig. 4. Tables 4 and 5. give the radiance at the entrance pupil at latitude 30° in north hemispheres with the ground reflectance 0.05 and 0.65 before and after the LTDN drift, units is $W/(m^2.Sr.\mu m)$.

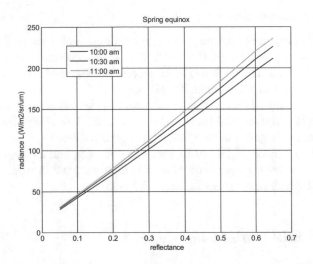

Fig. 4. The radiance at the entrance pupil of different LTDN at the same latitude region with the ground reflectance ranges from 0.05 to 0.65 in spring equinox

5.2   Imaging SNR Analysis of Remote Sensor

The variation of the radiance at the entrance pupil has influence on the remote sensor's imaging SNR.

Signal-to-noise ratio (hereinafter named as SNR) is one of the key parameters to quantitatively evaluate the image quality and radiometric performance of a space remote sensor. SNR reflects the radiometric resolution of a remote sensing system, as well as the level of scene and definition. SNR is defined as the ratio of signal output and root-mean-square noise [7]:

$$SNR = 20\log_{10}\left(\frac{s}{n}\right)$$

Where s is the signal output of the remote sensor, and n is the root-mean-square noise. And the electrons of signal [7] are:

$$s = E_s = \frac{\pi}{4}\left(\frac{1}{F}\right)^2 (1-\varepsilon)t_{int}N \int_{\lambda_1}^{\lambda_2} \tau(\lambda)R(\lambda)L(\lambda)\,d\lambda$$

Table 4. The radiance at the entrance pupil with different LTDN of ground reflectivity 0.05 at latitude 30° in north hemisphere (units: $W/(m^2.Sr.\mu m)$)

LTDN	Observation time			
	Spring equinox	Summer solstice	Autumnal equinox	Winter solstice
10:00 am	27.80	32.37	28.00	20.18
10:30 am	29.22	33.96	29.32	21.32
11:00 am	30.31	35.18	30.27	22.10

Table 5. The radiance at the entrance pupil with different LTDN of ground reflectivity 0.65 at latitude 30° in north hemisphere (units: $W/(m^2.Sr.\mu m)$)

LTDN	Observation time			
	Spring equinox	Summer solstice	Autumnal equinox	Winter solstice
10:00 am	211.26	253.72	213.15	131.57
10:30 am	224.76	267.31	225.71	143.84
11:00 am	235.13	276.13	234.81	152.17

Where F is the relative aperture, ε is the sheltering coefficient, t_{int} is the integration time, N is the stage, τ is the transmittance of the system, L is the radiance, and R is the responsivity of the detector. R can be defined by $R(\lambda) = \frac{\lambda}{hc} \cdot A \cdot FF \cdot \eta(\lambda)$, where h is Plank constant, c is velocity of light, A is the area of single pixel, FF is the fill factor of the detector, and η is the quantum efficiency.

With the CDS technology applying in the camera system, the noise of camera is mainly photon shot noise, detector noise, quantization noise, and electro-circuit noise. The noise of camera can be expressed as

$$n = \sqrt{n_{shot}^2 + n_{detector}^2 + n_{quantization}^2 + n_{electro-circuit}^2}$$

Panchromatic spectrum of the camera is covering 500 nm to 800 nm, F is 10, the spectrum transmittance is about 0.85, and the sheltering coefficient equals 0.16. 8192-pixels TDICCD of DALSA is selected as the detector of camera: the pixel pitch of detector is 7 μm, the fill factor of detector is approximately 89%, and stage 8 is selected for calculating in the simulation. The detector has an expected noise of 37 electrons. Output quantization is 10 bit. Electro-circuit noise of the camera can be controlled at 1.0 mV in engineering.

Through SAA and the radiance at the entrance pupil calculated before, the remote sensing system's imaging SNR with LTDN drift in different seasons was simulated and analyzed in this paper. The SNR change before and after the LTDN drift of 30 min with ground reflectivity 0.05 and 0.65 has been figured in Fig. 5, and the results of 4 typical solar terms at latitude 30° in north hemisphere are shown below (Tables 6 and 7).

The tables and figures above can show that imaging SNR varies with the LTDN drift. It is concluded that the imaging SNR decreases when LTDN drifted from 10:30 am to 10:00 am; and the imaging SNR increases when LTDN drifted from 10:30 am to

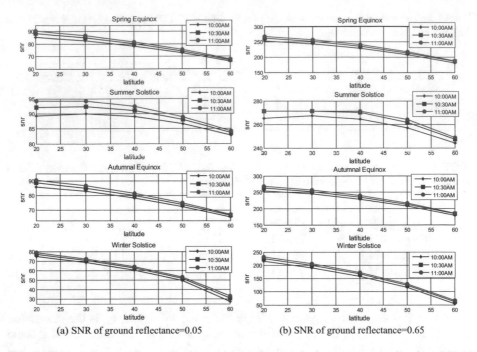

(a) SNR of ground reflectance=0.05   (b) SNR of ground reflectance=0.65

Fig. 5. SNR variation of 4 typical solar terms at different latitudes regions resulting from LTDN drift

Table 6. SNR with different LTDN of reflectivity 0.05 at latitude 30° in north hemisphere (units: dB)

LTDN	Observation time			
	Spring equinox	Summer solstice	Autumnal equinox	Winter solstice
10:00 am	38.33	39.08	38.36	36.70
10:30 am	38.57	39.31	38.59	36.98
11:00 am	38.75	39.48	38.75	37.17

Table 7. SNR with different LTDN of reflectivity 0.65 at latitude 30° in north hemisphere (units: dB)

LTDN	Observation time			
	Spring equinox	Summer solstice	Autumnal equinox	Winter solstice
10:00 am	47.72	48.54	47.76	45.61
10:30 am	48.00	48.67	48.02	46.01
11:00 am	48.20	48.67	48.19	46.26

11:00 am. The biggest variation of SNR with LTDN drift is: when LTDN drifted from 10:30 am to 10:00 am with ground reflectivity 0.65 at latitude 60° in north hemisphere, SNR reduces 0.93 dB, and a 9.16% decline.

6 Conclusion and Discussion

Through satellite orbit simulation, atmospheric radiation transmission simulation, and imaging signal-to-ratio calculation model, the variation of important imaging characteristics, which are important to the TDICCD camera's imaging quality, about imaging SNR, observation period and imaging coverage latitude range resulting from LTDN drift of ± 30 min from 10:30 am are simulated, calculated and analyzed in this paper. The analysis results show that the LTDN drift of sun synchronous orbit affects the remote sensing imagery quality.

In the ± 30 min of LTDN drift from 10:30 am of north hemispheres, SAA varies, sequentially affects the camera imaging, mainly about:

(1) Firstly, under the condition of SAA > 20° in the observation requirement, camera can imaging all year at the mid-low latitude (latitude < 42°); observation period has a change at the mid-high latitude (latitude > 42°), it decreases 5–15 days when the LTDN drifts from 10:30 am to 10:00 am, and increases 3–10 days when the LTDN drifts from 10:30 am to 11:00 am.

(2) Secondly, in the four seasons of a year, except that imaging coverage can reach the latitude > 85° in summer solstice without the influence of LTDN drift, it decreases nearly 0.6°–2.2° when the LTDN drifts from 10:30 am to 10:00 am, and increases nearly 0.1°–1.1° when the LTDN drifts from 10:30 am to 11:00 am.

(3) Lastly, when the LTDN drifts from 10:30 am to 10:00 am, the radiance at the entrance pupil decreases, the maximum reduction of imaging SNR can reach 0.93 dB, a 9.16% decline.

The imaging observation period, latitude observation scope and imaging SNR have been increased and optimized with the LTDN's shift towards positive values in north hemispheres. And reduction has been met with the LTDN's shift towards negative values in north hemispheres. This fact has influence on some satellites observing mission with high precision imaging requirement. Therefore, higher constraints and requirements of control on the LTDN's drift are needed.

Study and analysis result in this paper can give a quantitative reference to the visible light TDICCD camera's imaging quality analysis, then help designing more optimized on-orbit intelligent setup and adjustment strategy of imaging parameters. Meanwhile, it can provide a data support to the optimal design orbit and the shift control strategy of LTDN in future remote sensing satellite system.

References

1. Lillesand, T.M., Kiefer, R.W.: Remote Sensing and Image Interpretation. Publishing House of Electronics Industry, Beijing (2003)
2. Zhang, Y.-S., Feng, Z.-K., Shi, D.: The influence of satellite observation direction on remote sensing image. J. Remote Sens. Beijing **11**, 433 (2007)
3. Liu, L.: Orbit Theory of Spacecraft. National Defense Industry Press, Beijing (2001)
4. Zhang, G.-Y., Cai, L.-F., Huang, X.-F., Yang, Z.: Impact Analysis and Adjustment of Nearly Round Sun-Synchronous Orbit Satellite Injection Inclination Deviation. Aerospace Shanghai, Shanghai (2014)
5. Liu, X.F.: Basal Astronomy. Higher Education Press, Beijing (2004)
6. Hu, Z.W.: Common Astronomy. Nanjing University Press, Nanjing (2003)
7. Ma, W.P.: Space Optical Remote Sensing Technology, 1st edn, pp. 143–145. China Science and Technology Press, Beijing (2011)
8. Eric, F.V., Didier, T., Jean, L.T., et al.: Second simulation of the satellite signal in the solar spectrum, 6S: an overview. IEEE Trans. Geosci. Remote Sens. **35**(3), 675–676 (1997)

The Design and Application of Temperature Control Loop Heat Pipe for Space CCD Camera

Teng Gao[✉], Tao Yang, Shi-lei Zhao, and Qing-liang Meng

Key Laboratory for Advanced Optical Remote Sensing Technology of Beijing,
Beijing Institute of Space Mechanics and Electricity, Beijing 100094, China
gaoteng050125@163.com

Abstract. With the resolution of space CCD camera enhancing, the demand of temperature stability of CCD components becomes tight. The traditional thermal products, such as heat pipe, could not satisfy the temperature control precision. A Temperature-controlling loop heat pipe (TCLHP) using silicon nitride capillary wick is designed and produced successfully, which is used for thermal control of CCD components in satellite GF-9. Flight data indicated that the temperature precision was better than ±0.4 °C. It was the first time for Chinese home-grown loop heat pipe used for the thermal control of spacecraft in microgravity conditions, of which design method and flight data could provide a reference for the use of loop heat pipe in space.

Keywords: Loop heat pipe · Thermal control · Space CCD camera
High temperature precision · Design and application

1 Introduction

CCD component is the most core electrooptical device for space CCD camera, whose temperature level affects the image quality of camera. With the resolution of space CCD camera enhancing, the temperature stability of CCD component will further increase [1, 2]. CCD component has the features of small heat capacity, and high heat flux during periodic mode. For the camera focal plane mosaicked by multi CCD components, the difficulty of thermal control is large. Since the space for thermal control device is limited, micro/small heat pipes (the height of cross section is less than or equal to 5 mm) are used to transfer the heat produced by CCD components to the heat-collection plate, and then the heat on the heat-collection plate is transferred by large heat pipe and dissipated by radiators. However, this method does not meet the thermal control requirements of high precision due to small heat transfer capability or large heat transfer resistance [3, 4].

Loop heat pipe (LHP) is an advanced heat transfer device, which uses the capillary pressure generated in a porous structure to circulate the working fluid from heat source to heat sink in flexible pipes. Flexible pipes can be used on the rotating or mobile heat source. LHP has the characteristics of small heat transfer temperature difference, large heat transfer capability, long heat transfer distance and flexible piping layout. LHP is an

© Springer Nature Switzerland AG 2018
H. P. Urbach and Q. Yu (Eds.): ISSOIA 2017, SPPHY 209, pp. 65–74, 2018.
https://doi.org/10.1007/978-3-319-96707-3_8

advanced and promising thermal control device for spacecraft. As one upgraded product of traditional LHP, the evaporator of TCLHP and heat source are decoupled and cold plates are used to take away the heat produced by CCD components, as shown in Fig. 1. Cold plate component has multi evaporating units, which are connected in series or parallel by small diameter pipes. TCLHP has the characteristics of flexible layout and high temperature control accuracy. The working temperature of CCD components can be controlled by stabilizing the temperature of accumulator. TCLHP is preferable to be used in the thermal control of dispersed heat source for spacecraft [5–7].

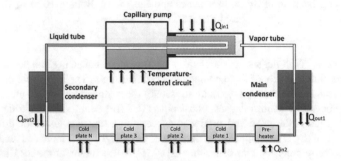

Fig. 1 The schematic of TCLHP

LHP was invented by Yu.F. Maydanik in Russia in 1972 [8]. The flight experiments were conducted in 1989 aboard the Russian spacecrafts "Gorizont" and "GRANAT". The behaviors of start-up and stable operation in conditions of microgravity were verified [9, 10]. The first actual application of LHPs took place in 1994 aboard the Russian satellite "OBZOR", where two LHPs were installed in the thermoregulation system of the unit of optical instruments [11]. In 1997, LHP was tested in space environment aboard the American shuttles "STS-87", "STS-83" and "STS-94", which verified the operating ability in orbit for long time [12, 13]. At present LHPs are widely used in space, where several tens of such devices are already successfully functioning. On American spacecraft ICESat, launched in 2003, two propylene LHPs was used for cooling the Geoscience Laser Altimetry System (GLAS) [14]. In the same year, eight LHPs were used for the temperature control of nickel hydrogen battery of Russian spacecrafts "YAMAL-200" [15]. Two LHPs transported waste heat of 208 W of 256 detector modules for the Burst Alert Telescope onboard NASA's Swift spacecraft in 2004 [16]. Six LHPs imported from Russia were used in the cooling system of nickel-cadmium batteries on a Chinese meteorological satellite FY-1 in 1999 [17].

In September 2015, a TCLHP was installed on the satellite GF-9, it was the first time for home-grown LHP in China being used for the thermal control of CCD components. The temperature precision of ±0.4 °C was realized. This paper briefly introduces the technical conditions of LHP first, and then the orbital flight data were analyzed, which will provide a reference for LHP used in conditions of microgravity for the thermal control of spacecraft.

2 Descriptions of TCLHP

The working fluid of TCLHP is ammonia. TCLHP is composed by capillary pump, cold plate, condensers, vapor and liquid tubes. Brazing is used for the connection of these components. The capillary pump consists of evaporator, accumulator and capillary wicks. Stainless steel tubes of 3 mm outer diameter are used for the vapor and liquid lines. Cold plate components are composed by four CCD cold plates which are coupled with CCD components, one preheater and connection pipes, as shown in Fig. 2. Condensers components consist of main and secondary condenser and pipes. Pipes are coupled with condensers by vacuum brazing. Table 1 shows the main parameters of TCLHP's each component.

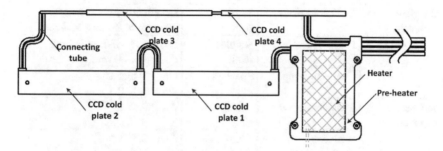

Fig. 2 Schematic diagram of CCD cold plates component

3 Design of Thermal Control Components

3.1 Thermal Physical Model

By neglecting the heat exchange between TCLHP and environment and sensible heat of superheat vapor, the energy balance equations of TCLHP and its sections can be written as:

$$Q_E + Q_{CC} + Q_{HC} + Q_{CCD} + \dot{Q}_{Ext1} + \dot{Q}_{Ext2} = \varepsilon\sigma A_{Con1}T_{Con1}^4 + \varepsilon\sigma A_{Con2}T_{Con2}^4 \quad (1)$$

$$\varepsilon\sigma A_{Con1}T_{Con1}^4 = \dot{Q}_{Ext1} + \dot{q}_m h_e + \dot{q}_m c_p(T_{cc} - T_{l2}) \quad (2)$$

$$\varepsilon\sigma A_{Con2}T_{Con2}^4 = \dot{Q}_{Ext2} + Q_{CCD} + Q_{HC} + \dot{q}_m c_p(T_{l2} - T_{l1}) \quad (3)$$

$$Q_E = \dot{q}_m h_e + Q_{Leak} \quad (4)$$

$$Q_{CC} = \dot{q}_m c_p(T_{cc} - T_{l1}) - Q_{Leak} \quad (5)$$

where Q_E is the driving power loaded on the evaporator; Q_{CC} is the temperature control power loaded on the accumulator; Q_{HC} is the compensation power on the pre-heater; Q_{CCD} is the power dissipation of CCD components; $\dot{Q}_{Ext1}$ is the external heat flux of

Table 1 Parameters of TCLHP

Components		Materials	Parameters
Capillary pump	Evaporator	Stainless steel	Size: Φ19 mm × 125 mm
	Accumulator	Stainless steel	Size: Φ36 mm × 140 mm
	Capillary wick	Silicon Nitride	Pore radius/Porosity: 1 μm/65%
			OD × length: Φ14 mm × 200 mm
Vapor and liquid pipes	Vapor pipe	Stainless steel	OD × Length: Φ3 mm × 480 mm
	Liquid pipe	Stainless steel	OD × Length: Φ3 mm × 480 mm
Cold plate component	CCD cold plate	Stainless steel	Length × width × height: 108 mm × 20 mm × 4 mm
	Pre-heater	Stainless steel	Length × width × height: 80 mm × 50 mm × 4 mm
	Connecting pipe	Stainless steel	OD × Thickness: Φ2 mm × 0.4 mm
Condenser component	Main condenser	Aluminum alloy	Area: 0.35 m^2
	secondary condenser	Aluminum alloy	Area: 0.33 m^2
	Main condenser pipe	Nickel	OD × length: Φ3 mm × 4900 mm
	Secondary condenser pipe		OD × length: Φ3 mm × 3250 mm
Working fluid		Ammonia	73 g

main condenser; $\dot{Q}_{Ext2}$ is the external heat flux of secondary condenser; ε is the emissivity of condensers; σ is Stefan–Boltzmann constant, 5.67×10^{-8} W/(m$^2 \cdot$ K^4); A_{Con1} is the area of main condenser; A_{Con2} is the area of secondary condenser; T_{Con1} is the effective temperature of main condenser; T_{Con2} is the effective temperature of secondary condenser; $\dot{q}_m$ is the mass flux; c_p is the specific heat at constant pressure; T_{cc} is the temperature of accumulator; T_{l1} is the temperature of liquid return pipe into accumulator; T_{l2} is the temperature of liquid return pipe into pre-heater; h_e is the latent heat; Q_{Leak} is the heat leakage from evaporator to accumulator.

In order to realize the temperature control function, the above relations should meet at any time.

3.1.1 The Requirement of Driving Power Loaded on Evaporator

The driving force of whole loop system is generated by loading heat on the evaporator. Except for the back-conduction power from evaporator to accumulator (Q_{Leak}), the other heat (effective power) is used to evaporate the working fluid. In order to assure the fluid in cold plate components being in two phase condition, the effective power should be larger than the total power loaded on the CCD cold plate component, which power used to make the working fluid phase-change. The driving power loaded on the evaporator can be written as:

$$Q_E \geq Q_{CCD} + Q_{Leak} + Q_{HC} - \dot{q}_m c_p (T_{cc} - T_{l2}) \tag{6}$$

3.1.2   The Temperature Control Requirement for Accumulator

The accumulator's temperature decides the phase-change temperature of fluid in CCD cold plates. The high temperature control accuracy of CCD components can be realized by controlling the temperature of accumulator precisely. The back-conduction power from evaporator to accumulator and subcooling temperature of return liquid affect the temperature level of accumulator. The active temperature control method is used to control the accumulator's temperature. Hence, the subcooling power of return fluid should be larger than the heat leakage power, as shown by

$$\dot{q}_m c_p (T_{cc} - T_{l1}) \geq Q_{Leak} \tag{7}$$

3.1.3   The Requirement of Compensation Power Loaded on The Pre-heater

Before entering into the four CCD cold plates, the fluid in pre-heater will be changed to be saturated by loading heat on it. Then the saturated fluid is forced into the cold plates and heat produced by CCD components is absorbed by the latent heat while the temperature keeps near constant. Hence, the compensation power loaded on the pre-heater should be larger than the subcooling power of return fluid, as shown by

$$Q_{HC} \geq \dot{q}_m c_p (T_{cc} - T_{l2}) \tag{8}$$

Besides, in order to prevent the working fluid freezing in the operating condition of extreme low temperature, active temperature control powers are loaded on the main and secondary condensers.

3.2   The Design Results of Temperature Control Circuits

There are four CCD components on the camera focal plane. The maximum power dissipation of single CCD component is 9 W, and the total heat power is 36 W. The orbital period is 90 min and the operation period is 15 min. The temperature stability and uniformity of the four CCD components should be better than ±2 °C. The power loaded on each parts of TCLHP and the areas of condensers can be decided by the energy balance equations in the upper segment and Thermal Desktop software. Table 2 shows the design power and threshold values of temperature control circuits.

The on-off of heating circuits are controlled by the temperature controller. Ratio switch is used for temperature control and the temperature control period is six seconds.

Table 2 The design conditions of temperature control circuits

Components	Heating circuits		
	Circuits names	Power	Threshold values
Evaporator	Driving heating circuit	50 W	30 °C–31 °C
Accumulator	Temperature control circuit	10 W	3 °C–7 °C
Pre-heater	Compensation heating circuit	10 W	30 °C–31 °C
Main condenser	Antifreeze heating circuit	10 W	−41 °C to −40 °C
Secondary condenser	Antifreeze heating circuit	10 W	−41 °C to −40 °C

4   Flight Data in Orbit

4.1   Start-Up Procedure

When the satellite comes into orbit and the TCLHP has not started up, only the heating circuits of accumulator and antifreeze heating circuits of condensers are open, and other heating circuits are all closed. The start-up of TCLHP needs the time of two orbital periods. During the first orbital period, the compensation heating circuit on the pre-heater is open, and the fluid in which will be changed into vapor, which will push liquid into the capillary wick and wet it to benefit TCLHP start-up in orbit. During the second orbital period, the driving heating circuit on the evaporator is turn on. When the temperature difference between evaporator and accumulator reaches to a certain value (i.e. superheat degree), the capillary force will drive the working fluid circulation in the loop.

Figure 3 shows the temperatures of TCLHP during the first orbital period. Before the compensation heating power is loaded on the pre-heater, the temperature of accumulator is near 6.5 °C. The temperature of evaporator is −4 °C and is smaller than temperature of accumulator, which indicates the fluid in evaporator is in subcooling condition. After the compensation power is loaded on the pre-heater, the temperatures of evaporator and accumulator fall by 1 °C–2 °C, which shows that subcooling liquid in tubes enters into the capillary pump and the fluid in the capillary wick is supplemented further. The temperature of condensers is higher than −45 °C during the whole process.

Figure 4 shows the temperatures of TCLHP during the second orbital period. Before the driving power is loaded on the evaporator, the temperature of evaporator is near −7 °C and the temperature of four CCD cold plates is in the range of 3 °C–6 °C. After the driving power is loaded on the evaporator, of which the temperature increases quickly. After 128 s, the evaporator's temperature reaches at 7.4 °C, which is higher 1.5 °C than the accumulator's temperature. The temperature of evaporator stabilizes gradually, and then the temperature of pre-heater decreases sharply from 30 °C to 7 °C. The temperature of four CCD cold plates is in the range of 5.8 °C–6.2 °C. These several phenomena indicate that TCLHP starts up successfully.

As heat is applied to the evaporator, of which the temperature rises almost perpendicularly. When the superheat reaches a certain value, nucleate boiling begins and working fluid vaporizes on the surface of the wick. Capillary pressure is generated in a

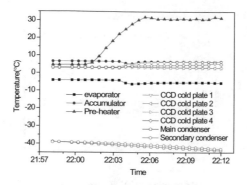

Fig. 3 The temperatures of TCLHP during the first orbital period

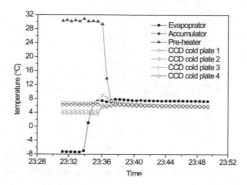

Fig. 4 The temperatures of TCLHP during the second orbital period

porous structure to circulate the working fluid in the loop. Vapor condenses into liquid in the main condenser and the capillary force continues to push liquid into the pre-heater, in which the fluid will be heated into saturated state. The pre-heater temperature keeps near 7 °C. Then the saturated fluid is forced into the four cold plates and heat produced by CCD components is absorbed by the latent heat of working fluid while the temperature keeps near 6 °C. Two-phase fluid condenses into subcooling liquid again in the secondary condenser and the capillary force pushes it into the accumulator. And this cycle repeats.

4.2 Stable Operation Procedure

Figure 5 shows the temperatures of TCLHP when camera is closed. The temperatures of evaporator and accumulator keep 7.1 °C and 5.6 °C, and the temperatures of four CCD cold plates are in the range of 5.5 °C–5.9 °C. Figure 6 shows the temperatures of TCLHP when camera is open. The turn-on time of camera is 11:06. The temperature of accumulator rises by 0.15 °C. The temperatures of evaporator and four CCD cold plates increase 0.4 °C–0.7 °C. The temperature uniformity of four CCD cold plates always keeps in the range of ±0.2 °C.

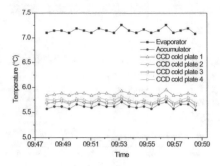

Fig. 5 The temperatures of TCLHP when camera is closed

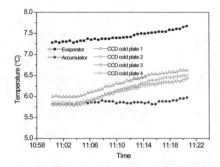

Fig. 6. The temperatures of TCLHP when camera is open

After camera is turned on, the quality of fluid in CCD cold plates increases. The flow resistance of whole system increases causing by the length of vapor phase in CCD cold plates and secondary condenser increases. The temperature of evaporator has to rise in order to overcome increased flow resistance. As a result, the phase-change temperature of fluid in CCD cold plates increases.

4.3   Longtime Temperature Comparison of CCD Cold Plates

Figure 7 shows the temperature comparison of CCD cold plates. The time of data in the left figure is from *July* 8th to *July* 10th, 2016. The temperature of four CCD cold plates is in the range of 5.7 °C–6.4 °C. The time of data in the right figure is from *July* 8th to *July* 10th, 2017. The temperature of CCD cold plates varies from 5.7 °C to 6.5 °C. The TCLHP of GF-9 satellite has worked continuously in orbit for almost two years. The temperature of four CCD cold plates is stable. The longtime temperature precision is better than ±0.4 °C.

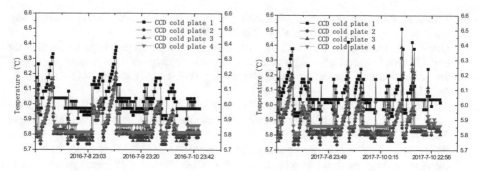

Fig. 7 Temperature Comparison of CCD Cold Plates in July 8th–10th between 2016 and 2017

5 Conclusions

In this paper, the TCLHP consisting of a silicon nitride capillary wick, stainless steel shell and working fluid of ammonia, which was launched in September 2015 on the GF-9 satellite, has the excellent microgravity adaptability and temperature control performance. It was the first time for Chinese home-grown LHP used for the thermal control of spacecraft. With the help of the TCLHP, ±0.4 °C of temperature control accuracy can be obtained for CCD components of GF-9 satellite in the condition of microgravity of space. Before start-up of the TCLHP, the compensation heating circuit on the pre-heater is turned on, in which the fluid will be heated into vapor, which will push liquid into the capillary wick and wet it to benefit start-up of TCLHP in orbit. This method increases the reliability of start-up for TCLHP in microgravity condition of space, which will provide a reference for TCLHP used in space.

References

1. Chen, E., Lu, E.: Thermal engineering design of CCD component of space remote-sensor. Opt. Precis. Eng. **8**, 522–525 (2000)
2. Cai, K., Wu, Q., Guo, J., Luo, Z., et al.: Thermal design of CCD focal plane assembly of space optical remote-sensor. Oftical Tech. **34**, 401–407 (2008)
3. Chen, S.: Some issues about space remote sensing. Spacecr. Recovery Remote Sens. **32**, 1–8 (2011)
4. Li, C.: Research on space optical remote sensor thermal control technique. J. Astronaut. **35**, 863–869 (2014)
5. Ding, T., Guo, L., Zhang H., et al.: The development status and future trend of heat pipe technology in space. In: Proceedings of 13th International Heat Pipe Conference, Shanghai, China (2004)
6. Zhao, Z., Lei, W.: The application of loop heat pipe in CCD thermal design. In: Proceedings of 12th International Heat Pipe Conference, Shenzhen, China (2010)
7. Cao, J.: The discussions of temperature control principal and method for loop heat pipe. In: Proceedings of 10th International Heat Pipe Conference, Guiyang, China (2006)
8. Maydanik, Y.: Loop heat pipes. Appl. Therm. Eng. **25**(5–6), 635–637 (2005)

9. Maidanik, Y.F., Fershtater, Y.G., Pastukhov, V.G., et al.: Thermoregulation of loops with capillary pumping for space use. SAE Paper No. 921169 (1992)
10. Orlov, A.A., Goncharov, K., Kotliarov, E.Y., et al.: The loop heat pipe experiment on board the GRANAT spacecraft. In: Proceedings of 6th European Symposium on Space Environmental and Control Systems, Noordwijk, The Netherlands, pp. 341–353, 20–22 May 1997
11. Goncharov, K., Nikitkin, M.N., Golovin, O., Fershtater, Y.G., Maidanik, Y.F., Piukov, S.A.: Loop heat pipes in thermal control systems for Obzor spacecraft. SAE Paper No. 951555 (1995)
12. Parker, M.L., Drolen, B.L., Ayyaswamy, P.S.: Flight test performance of a loop heat pipe – focus on a long steady state with no apparent subcooling. In: Proceedings of STAIF, Albuquerque, NM, pp. 818–823, January 31–February 4 1999
13. Yun, J.S.: American loop heat pipe (ALPHA) flight experiment. In: CPL 1998 International Workshop, March 2–3 1998
14. Grob, E.W., Baker, C.L., McCarthy, T.V.: Geoscience laser altimeter system (GLAS) loop heat pipes-an eventful first year on-orbit. SAE Paper No. 2004-01-2558 (2004)
15. Goncharov, K., Buz, V., Elchin, A., Prokhorov, Y., Surguchev, O.: Development of loop heat pipes for thermal control system of nickel – hydrogen batteries of "Yamal" satellite. In: Proceedings of the 13th International Heat Pipe Conference, China. China Astronautic Publishing House, Beijing (2004)
16. Choi, M.K.: Thermal assessment of Swift BAT instrument thermal control system in flight. SAE Paper No. 2005-01-3037 (2005)
17. Goncharov, K., Golovin, O., Kolesnikov, V., Xiaoxiang, Z.: Development and flight operation of LHP used for cooling nickel-cadmium batteries in Chinese meteorological satellites FY-1. In: Proceedings of the 13th International Heat Pipe Conference, Shanghai, China. China Astronautic Publishing House, Beijing (2004)

Integrated Thermal Design of One Space Optical Remote Sensor

Chunmei Shen[1,2(✉)] and Wenkai Liu[1,2]

[1] Beijing Institute of Space Mechanics and Electricity, Beijing, China
123855964@qq.com
[2] Key Laboratory for Advanced Optical Remote Sensing Technology of Beijing,
Beijing, China

Abstract. Thermal/electrical/structural integrated thermal design conception was applied for one space optical remote sensor. According to the integrated design conception, component structure and calorigenic equipments layout of remote sensor are optimized, and heat capacity and heat consumption of calorigenic equipments are managed systemically with the purpose of reducing heat pipe layout difficulty, decreasing number of heat pipe used for heat transfer, controlling remote sensor weight, and saving electrical power for heaters. Application of the integrated design conception on the remote sensor fully proved its advantages.

Keywords: Integrated thermal design conception · Thermal control
Space optical remote sensor

1 Introduction

Space optical remote sensor is very important payload of remote sensing satellite. It will suffer from a variant of adverse thermal environments, which include orbit external heat flux, cold black space background and other external thermal environment having great temperature fluctuation, such as satellite board [1, 2]. Meanwhile, under orbital vacuum conditions temperature of calorigenic equipments on remote sensor could be too high to burn out themselves or have an adverse effect on optical lens around them, if there is no any thermal control measures applied on them. Therefore, appropriate thermal control measures are necessary to make sure temperature of all components on remote sensor meeting requirements so that remote sensor can operate normally on orbit.

Resources, such as power for heating, weight, heat pipe layout space, radiator layout space and so on, are needed for thermal control of remote sensor. After years of development, mature idea of thermal control has been developed [1–7]. As long as thermal control resources are enough, mature idea of thermal control can almost meet all thermal control demands. However, with development of earth observation and remote sensing technology, thermal control engineer is facing more and more challenges, such as there are large number of calorigenic equipments, of which power consumption is very large and operating time is long, and sometimes the temperature level requirement of calorigenic equipments with large power consumption is very low. For example, the space spectral imager involved in [3] includes three video processors

© Springer Nature Switzerland AG 2018
H. P. Urbach and Q. Yu (Eds.): ISSOIA 2017, SPPHY 209, pp. 75–83, 2018.
https://doi.org/10.1007/978-3-319-96707-3_9

and two pulse tube cry coolers. The total peak heat consumption of these calorigenic equipments reaches 362 W. The maximum temperature requirement for the two pulse tube cry coolers, peak heat consumption of which is 295 W, has to be less than 10 °C. As another example, remote sensor in this paper includes seven electric devices in a small-scale space. And operating time of these electronic devices is long. Under such conditions, traditional thermal control methods need large amount of thermal control resources, such as larger radiating surface area, a large number of heat pipes, larger compensation power for calorigenic equipments under cold conditions, and so on. Nevertheless, resources available on orbit are limited. Then, new design idea should be developed to save thermal control resources as much as possible.

In this paper, design idea of integrating structure, electric and thermal, which focuses on saving thermal control resources, was applied on one remote sensor thermal design. Its advantages were fully proved.

2 Integrated Thermal Design Idea

The content of the integrated thermal design idea proposed in this paper includes electrical-thermal integrated design and structural -thermal integrated design.

Electrical-thermal integrated design is based on thermal management concept. It aims at managing the heat capacity and heat consumption of calorigenic equipments systematically. For calorigenic equipments, whose heat produced during operating must be dissipated under hot conditions, heat dissipation measures are designed around saving radiating surface area and cold condition compensation power. For calorigenic equipments, for which heat dissipation is not necessary under hot conditions, thermal control methods are designed to make use of their heat consumption as temperature control power for optical lens or as cold condition compensation power for calorigenic equipments needing heat dissipation. Therefore, the total temperature control power of thermal control system can be saved. For example, in [3], heat of electronic devices was designed to be used as temperature control power for optical lens successfully.

By using the electrical-thermal integrated design in this paper, radiating surface area can be saved under such conditions as follows: (1) Heat consumption of calorigenic equipment is used as temperature control power for optical lens. For this condition, Radiator is not needed anymore. (2) When calorigenic equipment, whose heat produced during operating must be dissipated under hot conditions, is connected to calorigenic equipment, whose operating temperature (for example 25 °C) is much lower than upper temperature limit (for example 45 °C) without cooling measure, if temperature of the latter equipment arises obviously (for example temperature of the latter equipment changes from 25 °C to 35 °C), radiator surface area of the former equipment can be saved. Because under this condition, part heat consumption of the former equipment is absorbed by the latter equipment, then the total heat needs to be dissipated by radiator is reduced. So radiator surface area can be saved. The benefit of saving cold condition compensation power is one of the results of radiator surface area reduction. Meanwhile, when calorigenic equipment, whose heat produced during operating must be dissipated under hot conditions, is connected to calorigenic equipment for which heat dissipation is not necessary under hot conditions, a system with lager heat capacity is composed by the two type calorigenic equipments. Then temperature decrement of the system must be smaller than that of the former calorigenic equipment itself during the same time period. So cold condition compensation power can be saved too.

Structural-thermal integrated design aims at making sure systematical management of calorigenic equipments fully guaranteed in structure, as well as decreasing difficulty of heat pipe layout, reducing the number of heat papers, and saving weight, through reasonable intervention in structure design from the angel of thermal control.

Electrical-thermal integrated design and structural -thermal integrated design complements each other, making the purpose of saving thermal control resources be achieved eventually.

3 Integrated Thermal Design of One Remote Sensor

3.1 Original Structure Layout of Remote Sensor

The remote sensor is composed of base plate, shell, earth baffle, optical lens, and seven electric devices. Figure 1 shows the remote sensor's initial structure layout. Optical lens and seven electric devices are all mounted on the base plate and all inside the shell. Seven electric devices are calorigenic equipments. Table 1 summarizes power consumption and working time of seven electric devices. It can be seen that total power of electric devices is large, and operating time electric devices is long. Figure 2 represents the payloads' deployment on satellite platform. There are three other payloads on satellite platform besides remote sensor.

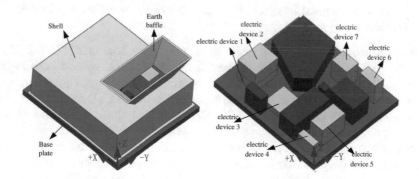

Fig. 1. Original structure layout of remote sensor

Table 1. Heat consumption and operating time of calorigenic equipments

Components	Heat consumption (W)	Operating time (min)	Temperature requirement (°C)
Electric device 1	20.1	63	−5 to 45
Electric device 2	35.0	68	
Electric device 3	3.0	102	
Electric device 4	13.0	14	
Electric device 5	35.5	63	
Electric dcvicc 6	8.0	102	
Electric device 7	5.0	51	

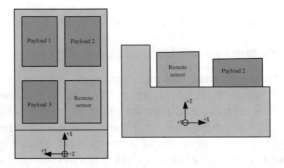

Fig. 2. Payloads' deployment on satellite platform

3.2   Electrical-Thermal Integrated Design Idea of Remote Sensor Electric Devices

Temperature requirement of base plate, shell and optical lens is 20 °C ± 2 °C, and −5 °C to 45 °C for seven electric devices. Software UG NX was adopted to create thermal analysis model of remote sensor. Electric devices operating temperature on orbit were calculated under the following boundary conditions: (1) temperature of base plate and shell is set to fixed value of 20 °C, (2) infrared emittance of base plate and shell inner surfaces is 0.85, as well as outer surfaces of seven electric devices, (3) outer surfaces of base plate and shell are covered with MLI, (4) no insulation between base plate and electric devices. The calculation results are shown in Table 2. It can be seen that peak temperature of electric device 1, 2, 5, 6 is around 65 °C, and 34 °C for electric device 3, 4, 7, which indicts that heat dissipation to cold black space background are necessary for electric device 1, 2, 5, 6, while not for electric device 3, 4, 7. Based on electrical-thermal integrated design idea, there are two methods to make use of heat consumption and heat capacity of electric device 3, 4, 7. One is that heat consumption of electric device 3, 4, 7 can be conducted to base plate or shell to save their thermal control power. But besides heat pipes between electric device and base plate and shell, heat pipes making heat conducted to base plate and shell from electric device distribute uniformly are also necessary. Otherwise, heat might concentrate at a small area and make temperature of this small area higher than temperature demand, which is 20 °C ± 2 °C. The second is that electric device 3, 4, 7 can be connected to electric device needing heat dissipation to compose a system with larger heat capacity, saving radiating surface area and cold condition compensation power. For the second method, heat pipes distributing heat uniformly are not needed. Because temperature requirement of electric devices is a large range of −5 °C to 45 °C. So the second method was utilized in this paper.

Table 2. Calorigenic equipments' operating temperature at room temperature environment on orbit

Components	Temperature (°C)
Electric device 1	61.3–67.5
Electric device 2	58.2–67.7
Electric device 3	32.2–32.6
Electric device 4	25.7–33.8
Electric device 5	59.8–70.1
Electric device 6	63.7–63.8
Electric device 7	24.5–34.5

Electrical-thermal integrated design idea of electric devices is described in detail as follows.

In original structure layout of remote sensor, electric device 2 is mounted on the top of electric device 1, and electric device 3 operating all the time during the whole orbit cycle is close to electric device 1(Fig. 1). So electric device 1, 2, 3 can compose a system having larger heat capacity and long-term heat consumption, by daubing thermally conductive filler on interface between electric device 1 and electric device 2, and connecting electric device 3 to electric device 1 thermally using heat pipe. Then heat dissipation measure is designed for the system.

In initial structure layout of remote sensor, electric device 4 is next door to electric device 5. Electric device 5 is next door to electric device 6. And electric device 6 operating all the time during the whole orbit cycle is next door to electric device 7. It is convenient to connect electric devices thermally next to each other using heat pipe. Therefore, electric devices 4, 5, 6, 7 can compose another system having larger heat capacity and long-term heat consumption. Then heat dissipation measure is designed for the system.

In conclusion, based on the electrical-thermal integrated design idea, original structure layout of remote sensor can be divided to three thermal control zones. Object temperature of two electric device system thermal control zone is −5 °C to 45 °C. And it is 20 °C ± 2 °C for optical lens and main structures thermal control zone. To protect temperature level of optical lens and main structures from being influenced by large temperature fluctuation of electric device, optical lens and main structures thermal control zone has to be insulated from electric devices thermal control zone effectively.

3.3   Structural-Thermal Integrated Design Under the Foundation of Electrical-Thermal Integrated Design

For remote sensor's original structure layout, through analysis, thermal control method for electric device 1, 2, 3 is given in Fig. 3a. That is putting radiator with type of converse L at the +X +Y corner outside of remote sensor, using four heat pipes going through shell to connect electric device 1 and 2 to +X side and +Z side of radiator respectively. Inside remote sensor, thermally conductive filler is used on the interface between electric device 1 and 2 to enhance conductive heat transfer between them. Electric device 3 is connected to electric device 1 using one heat pipe. To protect constant temperature zone of optical lens from being influenced by electric devices zone as much as possible, thermal insulation installation method is applied between base plate and electric device 1 and 3. And outer surfaces of electric device 1, 2, 3 are covered with MLI.

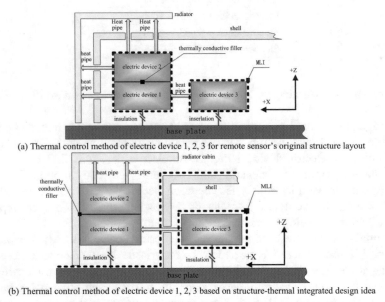

(a) Thermal control method of electric device 1, 2, 3 for remote sensor's original structure layout

(b) Thermal control method of electric device 1, 2, 3 based on structure-thermal integrated design idea

Fig. 3. Thermal control method of electric device 1, 2, 3

However, based on structure-thermal integrated design idea, thermal control method shown in Fig. 3b can be used for electric device 1, 2, 3. With the permission of the structure design, electric device 1 and 2 are moved to external of shell, the converse L type radiator is changed to a cabin, +X and +Z side of which are used as radiator. Electric device 1 and 2 are installed directly to +X side of the cabin, using thermally conductive filler on the interface. Electric device 2 is connected to +Z side of the cabin using two heat pipes. Electric device 3 is connected to electric device 1 using one heat pipe. Thermal insulation installation method is also applied between base plate and electric device 1, 3. Base plate and shell's surfaces facing to inner surface of radiator cabin and outer surfaces of electric device 3 are covered with MLI.

Compared to method shown in Fig. 3a, method shown in Fig. 3b has advantages as below:

(1) Number of heat pipe needing to go through shell is reduced to 1 from 4. Also, this makes heat pipe lay out design easier;
(2) The two heat pipes between electric device 1, 2 and +X side of radiator can be eliminated, deducing number of heat pipe, also reducing difficulty of heat pipe lay out design;
(3) To some extent, weight can be saved as a result of reduction of heat pipe and shell size, and elimination of traditional radiator support structures.

For remote sensor's original structure layout, through analysis, thermal control method for electric device 4, 5, 6, 7 is given in Fig. 4a. That is putting radiator with type of converse L at the −X +Y corner outside of remote sensor, using three heat pipes to connect electric device 4 and 5 to +Y side and +Z side of radiator respectively. Inside shell, electric device 4 and 5, electric device 5 and 6, electric device 6 and 7, are connected thermally with each other by heat pipe. Thermal insulation installation method is applied between base plate and electric device 4, 5, 6, 7. Outer surfaces of electric device 4, 5, 6, 7 are covered with MLI.

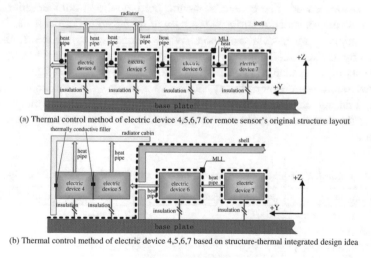

(a) Thermal control method of electric device 4,5,6,7 for remote sensor's original structure layout

(b) Thermal control method of electric device 4,5,6,7 based on structure-thermal integrated design idea

Fig. 4. Thermal control method of electric device 4,5,6,7

Based on structure-thermal integrated design idea, thermal control method shown in Fig. 4b can be used for electric device 4, 5, 6, 7. With the permission of the structure design, electric device 4 and 5 are moved to external of shell, and making sure electric device 4 and 5 contact to each other conductively through structure design. The converse L type radiator is changed to a cabin, +Y and +Z side of which are used as radiator. Electric device 4 is contacted conductively to +Y side of the cabin. +Z side of the cabin is connected to top side of electric device 4 and 5 using two heat pipes. Electric device 6 is connected to electric device 5 using one heat pipe going through shell. Inside shell, device 6 and 7 are still connected with each other by heat pipe. Thermal insulation installation method is still applied between base plate and electric device 4, 5, 6, 7. Base plate and shell surfaces facing to inner surfaces of radiator cabin and outer surfaces of electric device 6 and 7 are covered with MLI. Compared to method shown in Fig. 4a, method presented in Fig. 4b also has advantages of reducing heat pipe layout difficulty, saving heat pipe and weight, which is similar to method shown in Fig. 4b.

3.4   Design Results

Table 3 shows the operating temperature and standby mode temperature of the electric devices. Under standby mode (cold condition), heat consumption of device 1, 2, 4, 5, 7 is zero. Devices 3 and 6 have long term heat consumption. It can be seen from data in Table 3 that all the device temperature meets the temperature requirements. Operating temperature of device 3, 4, 7 are little higher than that shown in Table 2. This indicates that some heat of device 1, 2 produced during operating time is absorbed by device 3, and some heat of device 5, 6 is absorbed by device 4 and 7. Then heat of device 1, 2, 5, 6 needs to be dissipated by radiator is smaller than that in case that device 3 is not connected to device 1, 2, and device 4, 7 are not connected to device 5, 6. Then radiator surface areas are saved. The benefit of saving cold condition compensation power is one of the results of radiator surface area reduction. Meanwhile, device 1, 2 have no heat consumption under cold condition. When connected to device 3, the system composed by device 1, 2, 3 has long term heat consumption under cold condition. And heat capacity of the system composed by device 1, 2, 3 is larger than that of device 1, 2. Then temperature decrement of the system composed by 1, 2, 3 is smaller than that of device 1, 2 during the same time period. So cold condition compensation power can be saved. For the same reason, cold condition compensation power for device 4, 5 can be saved too.

Table 3. Calorigenic equipments' temperature on orbit by using integrated design method

Components	Operating temperature (°C)	Standby mode temperature (°C)
Electric device 1	28.5–38.3	−3.2 to −0.6
Electric device 2	27.9–38.1	−4.0 to −1.0
Electric device 3	33.2–36.9	−0.7 to 0.0
Electric device 4	28.2–34.1	−3.5 to −2.3
Electric device 5	29.1–36.8	−2.8 to −2.1
Electric device 6	27.1–30.8	2.8–5.3
Electric device 7	29.1–35.8	1.3–3.2

4   Conclusion

Design idea of integrating structure, electric and thermal was applied on one remote sensor thermal design. Based on thermal-electric integrated design idea, systematical management method of calorigenic equipments was presented. Then structure-thermal integrated idea was used to make systematical management of calorigenic equipments fully guaranteed in structure, through optimization of remote sensor structure and calorigenic equipments layout. Advantages of integrated design idea in saving thermal control resources were fully proved. Thermal design idea presented in this paper has important directive meaning to thermal design of remote sensor, especially for which includes large number of calorigenic equipments and operating time of calorigenic equipments is long.

References

1. Chen, Sh.P.: Space Camera Design and Test. China Astronautic Publishing House, Beijing (1991)
2. Min, G.R.: Satellite Thermal Control Technology. China Astronautic Publishing House, Beijing (1991)
3. Shen, Ch.M., Li, Ch.L., Gao, Ch.Ch.: Thermal management of Calorigenic Equipments in space spectral imager. Spacecr. Recovery Remote Sens. **33**, 80–85 (2012)
4. Yu, F., Xu, N.N., Zhao, Y., Xu, X.F., Feng, Y.G.: Thermal design and test for space camera on GF-4 satellite. Spacecr. Recovery Remote Sens. **37**, 72–79 (2016)
5. Zhao, Zh.M, Lu, P., Song, X.Y.: Thermal design and test for high resolution space camera on GF-2 satellite. Spacecr. Recovery Remote Sens. **37**, 34–40 (2015)
6. Chen, W.Ch., Wang, H.X.: Verification of thermal design and in-orbit flight for large off-axis triple-mirror anastigmatic camera. Opt. Instrum. **37**, 116–131 (2015)
7. Liu, W.Y., Ding, Y.L., Wu, Q.W.: Thermal analysis and design of the aerial camera's primary optical system components. Appl. Therm. Eng. **37**, 116–131 (2012)

Concept Design and Analysis for a High Resolution Optical Camera in Mars Exploration

Huadong Lian[✉], Weigang Wang, Wei Huang, Ruimin Fu, and Qingchun Zhang

Key Laboratory for Advanced Optical Remote Sensing Technology of Beijing, Beijing Institute of Space Mechanics and Electricity, Beijing 100094, People's Republic of China
lian_hd@163.com

Abstract. It is still very helpful and important to thoroughly investigate the mars surface by high-resolution optical remote sensor, although there have been some forthgoers such as the High Resolution Science Imaging Experiment (i.e. HiRISE) on Mars Reconnaissance Orbiter. Based on the requirements of the planned MARS exploration mission in China, a high resolution optical camera has been designed, manufactured and assembled in Beijing Institute of Space Mechanics and Electrics (BISME) recently. In the paper, the concept design about the camera was reported. The background of Mars observation was briefly overviewed firstly, especially for the high-resolution exploration. And then, the major technical requirements in the plan were listed for high resolution camera. In order to meet the requirements, it is necessary to keep the camera more light-weighted and more reliable. Therefore, several effective schemes were adopted such as the compacted optical system form, the advanced ultra-light-weighed materials, the optimization configuration and the integrated electrical device design, etc. The designs of the optical system, structural configuration and electrical scheme were introduced with some simulation results. A prototype camera was manufactured and tested. Experimental results showed that the camera had excellent performance and reliability which can satisfy the mission demands. It should be noted that with proper modification the optical camera can be applied in many other fields such as Earth observation.

Keywords: Mars exploration · High resolution optical camera
Light weighted

1 Introduction

As the nearest planet to earth, Mars has been observed from land to space and even landed by human kind for many years. Mars exploration activities are indeed helpful and important in many aspects, such as understanding the evolution history of the earth and solar system, forecasting the future of cosmos, and even searching for other terrestrial planets, etc. Therefore, more and more counties and organizations have been involved in the Mars exploration and research activities. Until the beginning of 2016,

© Springer Nature Switzerland AG 2018
H. P. Urbach and Q. Yu (Eds.): ISSOIA 2017, SPPHY 209, pp. 84–92, 2018.
https://doi.org/10.1007/978-3-319-96707-3_10

the total numbers of mars exploration had been up to 43 times. However the failure rate and risk were very high in the past. Thanks to the technical improvement in the 21 century, the mission success rate has risen greatly. In this time, many missions have been carried out, such as Spirit and Opportunity Rovers, Mars Reconnaissance Orbiter (MRO) by NASA [1]. By these missions, many beneficial achievements have been obtained, for example, the evidences of underground water in Mars. Although the considerable improvements have been acquired, many further exploration missions are still urgent and necessary to know the Mars thoroughly and completely.

High resolution optical camera has played very important roles in the Mars exploration, and there have been many such type cameras launched, such as High Resolution Stereo Camera (HRSC) on Mars Express [2], the high resolution Mars Orbital Camera (MOC) on Mars Global Surveyor [3] and the High Resolution Science Imaging Experiment (HiRISE) on Mars Reconnaissance Orbiter [4]. Figure 1 shows the configuration of the HiRISE [5]. The camera employed a 50 cm, f/24 all-reflective optical system and a time delay and integration (TDI) detector assembly to map the Mars surface from its orbital altitude ~ 300 km. HiRISE has three different colour bands from 0.4 to 1.0 μm, which can identify the mineral, ice and dust, etc., but not classify the different types of minerals. Form the past missions, one can see that the high resolution optical camera with several colour imaging modes is indispensable, especially for the identifying the detailed types of minerals, knowing the atmosphere environment, etc.

Fig. 1. The configuration of HiRISE camera

Many institutes and organizations in China have also involved in the Mars exploration activities recently. Yinghuo-1 was the first attempt to investigate the Mars on its orbit; however the mission was not successful due to the failure launch [6]. Now new program of Mars exploration missions in China have been arranged, in which the high resolution optical camera is undoubtedly one of the most important payloads. According to the requirements of high resolution optical camera in the plan, a camera was designed, simulated and verified in our institute, Beijing Institute of Space Mechanics and Electrics (BISME). In the paper, the concept design and initial analysis about the high resolution camera were introduced.

2 Technological Requirements on High Resolution Camera

In this section, several important technological requirements were introduced. The operating orbit has elliptical shape and some major specifications were listed as followed:

Semimajor axis	9502 km
Perigee altitude	265 km
Eccentricity	0.61
Inclination	93.1°
Period	7.8 h
Longitude of ascending node	115.9°
Perigee argument	117.6°
True anomaly	0°

The requirements on the high resolution optical camera are listed in the Table 1. From the table, one can see that the camera has four colour bands and one panchromatic band. The resolution should be less than 0.5 m and 10 m for pan and colour bands respectively. The weight and power should be less than 40 kg and 100 W respectively. It should be noted that only some basic specifications were listed here, and there were still many other requirements that should be satisfied in the design.

Table 1. Main performance requirements on the high resolution camera

Parameters	Requirement
Ground resolution (at altitude 265 km)	Pan: Less than 0.5 m in special region Colour: less than 10 m
Coverage (at altitude 265 km)	9 km
MTF (static, Nyquist)	>0.18
SNR (signal noise ratio)	>100
Wavelength bands	PAN: 0.45–0.52 Color: Blue 0.45–0.52 Green 0.52–0.60 Red 0.63–0.69 Near IR 0.76–0.90
Weight	Less than 40 kg
Power	Less than 100 watts

3 Concept Design of the High Resolution Camera

3.1 Requirement Parameters Analysis

From the performance requirements and the orbital parameters, one can see that it is difficult to finish the camera with traditional methods. Therefore, many new technologies were updated or developed in this task, such as the super light-weighted optomechanial structural design technique, high signal noise ratio (SNR) technique under low luminance condition and high image quality technique for large topological variation, etc. In the paper, emphasis was put on the super lightweighted design technique. In the following, some considerations were introduced to make the optical camera more lightweighted and more reliable.

Firstly, Korsch optical type with long focal length is apt to compact the optomechanical system, and many ultra-lightweight small camera have adapted such types (for example, the focal length of HiRISE is 24 m, and the gound resolution is 0.3 m@300 km). The F-number of the optical system defines the aperture size. With the F-number larger the aperture size becomes smaller when all the other parameters in optical system keep same. For the TDI detector, the F-number is not the key factor to determine the total energy; moreover it is still confined by the MTF values. By a trade balance, the F-number takes 10.5 and the focal length is about 3750 mm. The best resolution is about 0.49 m, better than the requested value. However, it should be noted that the resolution values are different in one orbital period due to the elliptical shape characteristic of the orbit. Only 19.2% and 32.2% of the arc length in one orbital period may have the better resolution for the pan and color bands respectively.

Secondly, adopting the suitable materials is crucial to reduce the weight, improving the stiffness and strengthening the thermal stability. The materials with low CTE (coefficient of thermal expansion), high elastic modulus to density ratio and strength to density ratio are certainly preferred. However it is difficult to implement in the engineering practice. In the design several rules were used for different parts. For the mirror and its support structures, thermal mismatch should be very small and both the reliability and quality of the optical figuring and alignment should be high. The mirrors here employed the ULE (ultra-low-expansion glass by Corning) material and three-point kinematic support structures. For the steering structure (tube), cyanate based CFRP, a material with extreme low CTE and density material made by BISME,was used to maintain the strict thickness tolerance between mirrors. All the materials should also have the excellent stability and reliability.

Thirdly, all the components and assemblies were designed from topological configuration to detailed dimension size in all possible operating conditions. Optomechanical integrated analysis by SORSA (Space Optical Remote Sensor Analysis platform in BISME, as showed in Fig. 2) was applied thoroughly in the design process to make the structure more lightweight and reliable.

Fourthly, the focal electrical plate equipment was designed by integrated scheme, thus producing a lightweighted and stable structural configuration. At the same time, the high reliability could be obtained because of backup design of electrical circuit.

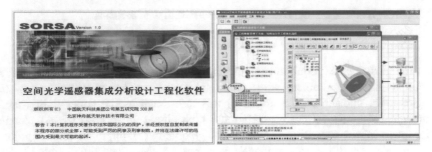

Fig. 2. The interface of the SORSA plateform

3.2   Optical System Design

Several different optical system designs were all considered and optimized by optical design software. The final optimal design takes the three mirror astigma (TMA) system which has a long focal distance and a large F number with medium field of view (FOV). Main parameters of the optical system were as followed:

Spectral bands	P: 0.45–0.90(PAN) B1: 0.45–0.52(BLUE) B2: 0.52–0.59(GREEN) B3: 0.63–0.69(RED) B4: 0.77–0.89(NIR)
FOV	$2.64° \times 0.25°$
Focal length	3750 mm
F-number	10.5
MTF	MTF $\geq$ 0.31 (@71.4 lp/mm (PAN) MTF $\geq$ 0.6 (@18 lp/mm (COLOR)
Transmissivity	≥ 0.75

The optical system configuration is given in Fig. 3. The primary mirror is the largest optical element with outer diameter 360 mm and inner diameter 100 mm. The secondary mirror has convex optical surface and the tertiary mirror has a rectangular pupil with dimension 160 mm × 56 mm. In this TMA optical form, two fold mirrors are valuable to keep the optical system more compact. The designed MTF values were 0.316 and 0.7 for PAN and color bands respectively in all fields, larger than the required value 0.31 and 0.5.

3.3   Structural Design and Configuration

From the optical system design and its detailed requirements, the optomechanical design was optimized. The camera configuration was obtained as shown in Fig. 4, and the main assemblies were listed in Table 2. The envelope dimension is 960 mm × 530 mm × 630 mm, a little smaller than the required value 964 mm × 631 mm × 818 mm, and the weight without the electrical plate assembly is about 24 kg.

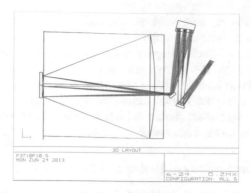

Fig. 3. Optical system design

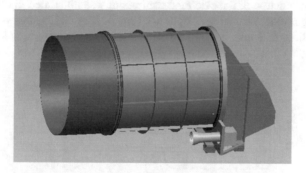

Fig. 4. Optomechanical configuration of the camera

Table 2. Main assemblies of the camera

Assembly name	Sub-assemblies or components
Optical element	Primary mirror and its mount
	Secondary mirror and its mount
	Fold mirrors
	Tertiary mirror and its mount
Stray light contorl	Front baffle
	Secondary mirror cover
	Primary mirror cover
Focal plane	Focal plane assembly
	Pixel velocity measuring element
Thermal control	Thermal pipe
	Radiating surface

The base frame, manufactured by C_f-SiC, has serval interfaces with spacecraft and the front steering structure. And at the same time, the focal plane assembly and aft optical elements are all mounted on the base frame. The tube was manufactured by cyanate based CFRP which has high strength and thermal stability.

All the optical elements have employed the ULE material and three-point kinematic mount. To attenuate the vibration impact on the mirror figure on orbit and the maximum stress level during launch, damping alloy was imposed on the three struts for the primary mirror assembly. The mirrors are bonded with the struts by inserts, and the insert employs the invar material which has the matching CTE with the mirror.

It also should be noted that a small transmissive optical system can be alternative to improve the ground resolution further, which is called the pixel-velocity measuring element and mounted on the base frame.

3.4   Electrical Plate Equipment Design and Consideration

The electrical plate equipment was designed in ultra-compact and integrated pattern, which included two parts: focal plane circuit subsystem and information processing subsystem. The function of the electrical equipment was illustrated in the left of Fig. 5, and the configuration was in the right of Fig. 5. The dimension of the electrical equipment is about 350 mm × 300 mm × 100 mm, and its weight is less than 6.0 kg.

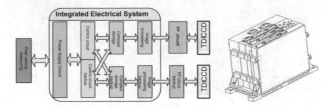

Fig. 5. The function and configuration of the integrated electrical equipment

4   Simulation Analysis and Verification

The performance specifications of the camera were all verified by simulation analysis and even prototype test, and some results were shown as followed. The finite element of the camera was given in the left of Fig. 6 and the constrained modal analysis was executed with the first modal contour in the right of Fig. 6. From the figure, one can see that the first frequency was beyond 114 Hz. Therefore it has great capacity to survive the launch vibration and maintain fine figure in the orbit environment.

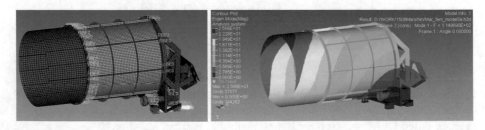

Fig. 6. Finite element model and 1st modal result

The stray light was also optimized and the model was given in Fig. 7. From the analysis, the influence of the stray light was decreased down to 3% by adding some shields, and the requirement of energy of the focal plane was also satisfied.

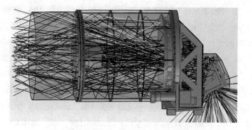

Fig. 7. Stray light analysis

The prototype was manufactured (as shown in Fig. 8), which survived the vibration and thermal vacuum tests and obtained fine images in the ground imaging test. Additionally, the weight was summarized in Table 3 for each assembly of the prototype.

Fig. 8. The prototype camera

Table 3. The weight for each assembly

Assembly name		weight(kg)	
Main Body Structure	Front sunshade	1	
	Secondary mirror Assembly	1.4	
	Tube Structure	3.8	
	Primary Mirror Assembly	4.6	
	Base Frame Structure	4.0	
	Fold Mirror Assembly 1	0.5	
	Tertiary mirror Assembly	1	24.
	Fold Mirror Assembly 2 and the FP Adjustment Assembly	1.5	
	FP Assembly	2	
	Thermal control	1.5	
	Electrical cable	0.3	
	Standard parts	0.4	
	Velocity measurement Unit	1.0	
Integrated Information Processing System		6.0	
Total Weight		31. kg	

5 Conclusion

In the paper, the concept design was introduced for a high resolution optical remote sensor in Mars exploration. The key idea in the design was to keep the camera more lightweight and more reliable. Many effective techniques were adopted, such as compacted optical system with large F-number, excellent optomechanical materials, integrated electrical equipment, etc. The whole weight of the camera was only 31 kg, less than the requested value, and the weights of all the assemblies were summarized in the following table. Some simulation and prototype test results were also listed in the paper to show the availability and reliability of the design. It should be also noted that with proper modification this camera can be applied in many other fields such as earth observation, and at the same time, by optimizing the combination of the design parameters, the ground resolution can be enhanced further.

References

1. Fogell, R.A., Meyer, M.A., McCuistion, J.D: NASA remote sensing plans for Mars exploration. In: Proceedings of SPIE, vol. 5978, pp. 59780B-1–59780B-12 (2005)
2. Schulster, J.R., Denis, M., Moorhouse, A.: From mission concept to mars orbit: exploiting operations concept flexibilities on Mars express. In: Space Optics Conference, vol. 5958, pp. 5958-01–5958-21 (2006)
3. Malin, M.C., et al.: Design and development of the mars observer camera. Int. J. Imaging Syst. Technol. 3(1), 76–91 (1991)
4. Ebben, T.H., Bergstrom, J., Spuhler, P., Delanmere, A., Gallagher, D.: Mission to Mars: the HiRISE camera on-board MRO. In: Proceedings of SPIE, vol. 6690, pp. 66900B-1–66900B-22 (2007)
5. Gallagher, D., Bergstrom, J., Day, J., Martin, B., Reed, T., Spuhler, P., et al.: Overview of the optical design and performance of the high resolustion science imaging experiment (HiRISE). In: Proceedings of SPIE, vol. 5874, pp. 58740K-1–58740K-10 (2005)
6. Xiong, W., Xie, C., Liang, X., et al.: Deep space TT&C equipments of YH-1. Chin. J. Space Sci. 29(5), 490–494 (2009). (in Chinese)

A Method of Vibration Suppression of a Space Mechanical Cryocooler Based on Momentum Compensation

Mingshuo Che$^{(\boxtimes)}$, Zhe Lin, and Wenpo Ma

Key Laboratory for Advanced Optical Remote Sensing Technology of Beijing,
Beijing Institute of Space Mechanics and Electricity, Beijing, China
cms1014@163.com

Abstract. The space mechanical cryocooler, which is one of the important parts to make sure the detect performance of remote sensors, is used to ensure that infrared detectors can be in stable and low temperature conditions in the space optical remote sensing system. However, a large vibration will be caused by the reciprocating motion of the compressor's piston of the mechanical cryocooler in principle. In this paper the control system of active vibration suppression of the space mechanical cryocooler was studied. This paper presented a control method based on the momentum compensation. The method is based on the dual piston compressor. The accelerometer, which is in direct proportion to the difference in momentum, is used to obtain the acceleration signal from the compressor shaft as a feedback signal for vibration. The momentum is offset between the slave piston and the master piston by compensating the driving voltage of the slave piston to suppress vibration. The two algorithms of PID and LMS are given in the part of parameters setting of the control method. Finally the control method is verified. The control method with both algorithms of PID and LMS in the part of parameters setting had fulfilled the general requirement for space infrared application and the process converged quickly.

Keywords: Space mechanical cryocoolers · Active vibration suppression
Adaptive control · Momentum compensation

1 Introduction

In the exploration of the universe and development of the earth's resources, remote sensing satellites have incomparable advantage over other carrier in terms of access to information. The infrared remote sensor is the key of the remote sensing satellite to complete the information collection equipment [1]. Modern infrared remote sensors usually choose infrared photon type semiconductor photoelectric detectors. The infrared semiconductor materials at high temperatures can produce inherent thermal excitation, resulting in large dark current and noise [2]. So in order to make the infrared detector to be in low temperature, space mechanical cryocoolers are needed. Space mechanical cryocoolers are primarily Stirling cryocoolers and pulse tube cryocoolers. Both the Stirling cryocooler and the pulse tube cryocooler, however, its piston recip-rocating motion will produce larger vibration in principle. This will affect the

© Springer Nature Switzerland AG 2018
H. P. Urbach and Q. Yu (Eds.): ISSOIA 2017, SPPHY 209, pp. 93–102, 2018.
https://doi.org/10.1007/978-3-319-96707-3_11

performance of instruments. For example, the vibration will lead to imaging blur. And on the other hand, the vibration will cause electromagnetic interference, causing a larger impact on some sensitive sensor [3].

The earliest cryocooler compressor adopts single piston compressor. But this kind of compressor output large vibration. In order to reduce the compressor vibration, two single piston compressors for opposite arrangement had been used to implement the momentum balance and vibration reduction. Later in order to reduce the volume and quality, a dual-piston compressor was born. The compressor of the cryocooler is currently used in this form such as that shown schematically in Fig. 1. This method, however, although producing quite acceptable vibration export for most applications, due to the dissimilarity of the opposing motors/pistons (different wear, friction factors, spring constants, clearance sealing, strength of permanent magnets, etc.) does not produce vibration free cryogenic coolers as needed in the above mentioned more demanding applications. Moreover, most of those asymmetry factors are runtime-dependent, thus in most cases a dual-piston compressor eventually looses its original in-factory balance and becomes "noisier" with time [4]. On the other hand, A large number of experiment studies showed that by using opposed-piston can only reduce the fundamental frequency vibration and the effect is limited for the high frequency vibration [5]. Therefore, it is of necessity that to take vibration reduction measures to meet the demand for low vibration of cryocoolers in some space application projects. As for vibration isolation and passive vibration reduction have some limitations, active vibration control method has become the dominant means to achieve higher performance of vibration suppression.

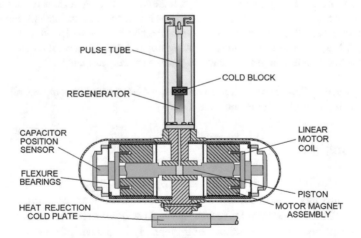

Fig. 1. Schematic of modern, dual-piston pulse tube space cryocooler

2 Active Vibration Control System

In this paper, the control method we studied is based on momentum compensation which is only using acceleration sensors in axial to get the acceleration signal as the feedback signal of vibration. The vibration signal is sampled synchronously and fed to a proprietary vibration control algorithm which continually adjusts the harmonic content of the two compressor drive waveforms. The two algorithms of PID and LMS are given in the part of parameters setting of the control method.

2.1 Fundamental Principle

The basic design of the control method relies on the dual-piston compressor. The two pistons of the compressor are defined as the "master piston" and "slave piston" respectively. The momentum is offset between the "master piston" and the "slave piston" by compensating the driving voltage of the "slave piston" to suppress vibration. This corrective voltage is comprised of the harmonics of the driving frequency, the magnitudes and phases of which are dynamically (real-time) adjusted with the purpose of minimizing the magnitudes of the respective harmonics of the measured vibration to the lowest possible level [6–8]. A block diagram of the active vibration control system is shown in Fig. 2.

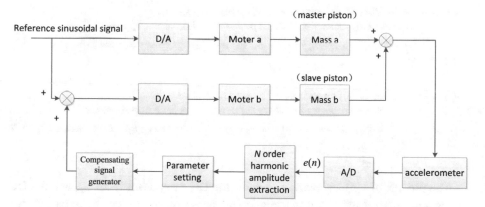

Fig. 2. A block diagram of the active vibration control system

In Fig. 2, high sensitivity accelerometer is used to measure the vibration of the compressor in axial. Measured acceleration went through the n order harmonic amplitude extraction part after the A/D converter. The n order harmonic amplitude extraction not only has the filter function, but also has the role of the detection of harmonic amplitude. With the n order harmonic, for example, the brief block is shown in Fig. 3.

After the acceleration signal changed into the digital signal from the analog signal, vibration signal $e(n)$ can be expressed in follow forms cause there are many high order harmonics besides the fundamental frequency.

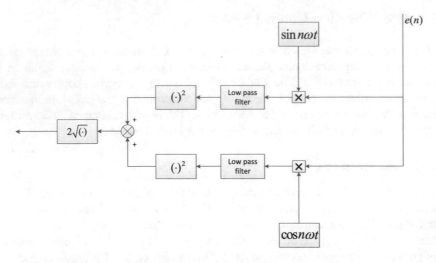

Fig. 3. Brief block of n order harmonic amplitude extraction

$$e(n) = A_1 \sin \omega t + A_2 \sin 2\omega t + \cdots + A_n \sin n\omega t \qquad (1)$$

$e(n)$ multiplied with $\sin n\omega t$ and $\cos n\omega t$ respectively, the product is given by

$$
\begin{aligned}
e(n) \cdot \sin n\omega t &= [A_1 \sin(\omega t + \phi_1) + A_2 \sin(2\omega t + \phi_2) + \cdots + A_n \sin(n\omega t + \phi_n)] \sin n\omega t \\
&= A_1 \sin(\omega t + \phi_1) \sin n\omega t + A_2 \sin(2\omega t + \phi_2) \sin n\omega t + \cdots + A_n \sin(n\omega t + \phi_n) \sin n\omega t \\
&= -\frac{A_1}{2}\{\cos[(1+n)\omega t + \phi_1] - \cos[(1-n)\omega t + \phi_1]\} - \frac{A_2}{2}\{\cos[(2+n)\omega t + \phi_2] - \cos[(2-n)\omega t + \phi_2]\} \\
&\quad - \cdots - \frac{A_n}{2}[\cos(2n\omega t + \phi_n) - \cos\phi_n]
\end{aligned}
\qquad (2)
$$

$$
\begin{aligned}
e(n) \cdot \cos n\omega t &= [A_1 \sin(\omega t + \phi_1) + A_2 \sin(2\omega t + \phi_2) + \cdots + A_n \sin(n\omega t + \phi_n)] \cos n\omega t \\
&= A_1 \sin(\omega t + \phi_1) \cos n\omega t + A_2 \sin(2\omega t + \phi_2) \cos n\omega t + \cdots + A_n \sin(n\omega t + \phi_n) \cos n\omega t \\
&= \frac{A_1}{2}\{\sin[(1+n)\omega t + \phi_1] + \sin[(1-n)\omega t + \phi_1]\} + \frac{A_2}{2}\{\sin[(2+n)\omega t + \phi_2] + \sin[(2-n)\omega t + \phi_2]\} \\
&\quad + \cdots + \frac{A_n}{2}[\sin(2n\omega t + \phi_n) + \sin\phi_n]
\end{aligned}
\qquad (3)
$$

The product of $e(n)$ and $\sin n\omega t$ contains the DC component $\frac{A_n}{2}\cos \phi_n$, and the DC component only related to the magnitude and phase of the n order harmonic. In the same way the product of $e(n)$ and $\cos n\omega t$ contains the DC component $\frac{A_n}{2}\sin \phi_n$, and the DC component only related to the magnitude and phase of the n order harmonic as well. Thus the magnitude of n order harmonic can be given by

$$A_n = 2\sqrt{\left(\frac{A_n}{2}\cos \phi_n\right)^2 + \left(\frac{A_n}{2}\sin \phi_n\right)^2} \qquad (4)$$

According to (4) the two DC components were separated from the two products respectively by low pass filters. Then with some basic operation the magnitude of the n order harmonic was obtained. The magnitudes of other high harmonics and the fundamental frequency can be obtained as well. Then the magnitude multiplied with the standard sine signal and cosine signal at this frequency after parameters setting. The sum of the two parts is the compensation signal at this frequency. Add all the compensation signals of the fundamental frequency and high harmonics together and then went through the D/A converter to drive the "slave piston" to get the equal momentum but opposite direction with the "master piston" to reduce the vibration of the compressor.

2.2 Parameters Setting

The two algorithms of PID and LMS are given in the part of parameters setting of the control method. In a certain harmonic, for example, the feedback is the intensity of the vibration signal. The goal of control is to make the feedback get to "0". PID or LMS can be used to solve the problem of parameters setting. Taking an example of PID, the brief block is shown in Fig. 4.

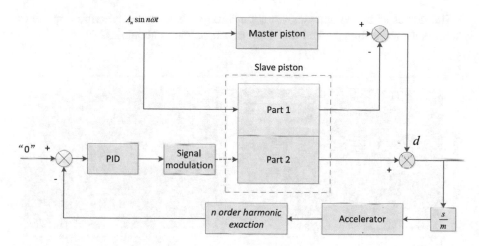

Fig. 4. Brief block of parameters setting of PID

The sine signal of n order frequency drives the master piston and the slave piston at the same time. The sum of the momentums of the two pistons is as the disturbance d. In order to make the control target tends to "0", the feedback loop is added in the system. Output e divided by the compressor casing quality m and then after the differential the acceleration is gotten. The magnitude of n order frequency is extracted after the acceleration has been measured by the accelerometer. Compare the magnitude with "0". The difference is the gap with the goal. Set the parameters by PID, and the compensation signal is generated through the signal modulation after parameters setting. Drive the "slave piston" by compensation signal. Then compare the compensation momentum with disturbance d to make the difference e trends to zero. The algorithm of LMS in the part of parameters setting is the same as PID.

3 Experiment

The summary of parameters of a cryocooler is shown in Table 1.

Table 1. Summary of parameters of a cryocooler

Parameter	Value
Motor resistance (Ω)	0.88
Motor inductance (H)	0.009
Motor force constant (N/A)	24
Viscous damping coefficient (Ns/m)	50
Total mass (kg)	10
Moving mass (kg)	1.14
Spring stiffness (N/m)	15000

The output of the compressor vibration force without the active vibration control system is shown in Fig. 5. And the value of the output is about 4.32 N.

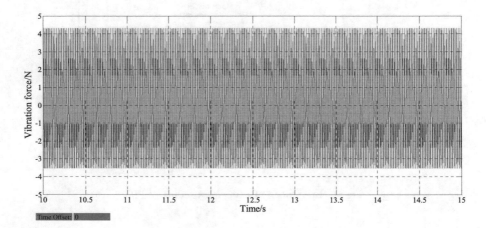

Fig. 5. Output of the vibration force without vibration control

According to the control method with two algorithms of PID and LMS in the part of parameters setting, the effect about the vibration suppression of the first three order frequencies of the compressor was simulated.

With control method with the algorithm of PID in the part of parameters setting, the output of the compressor vibration force in time domain is shown in Fig. 6 and the vibration spectrum is shown in Fig. 7.

Figure 6 shows that the vibration force was decreasing and the vibration force was reduced from 4.32 N to 0.01 N. The vibration force in the first three order frequencies was reduced 94%, 89% and 84% respectively as was shown in Fig. 7.

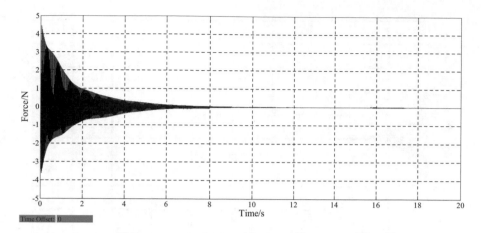

Fig. 6. Output of vibration force in time domain

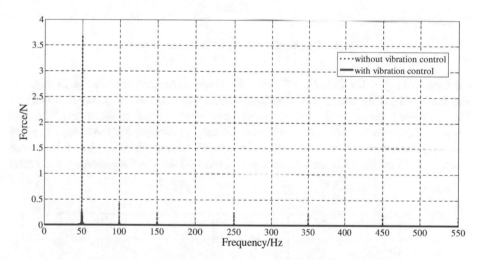

Fig. 7. The vibration spectrum of the output with and without vibration control

With control method with the algorithm of LMS in the part of parameters setting, the output of the compressor vibration force in time domain is shown in Fig. 8 and the vibration spectrum is shown in Fig. 9.

Figure 8 shows that the vibration force was decreasing and the vibration force was reduced from 4.32 N to 0.005 N. The vibration force in the first three order frequencies was reduced 96.4%, 89% and 94% respectively as was shown in Fig. 9.

Compared with two algorithms, the algorithm of PID had the advantages of simple calculation and convenient adjustment but it was not adaptable to the variation of model parameters. However, the algorithm of LMS had a good adaptability to the model but in the process of parametric convergence, if the initial coefficient was defined "0", the result was shown in Fig. 10, which converged and stable in a non-zero

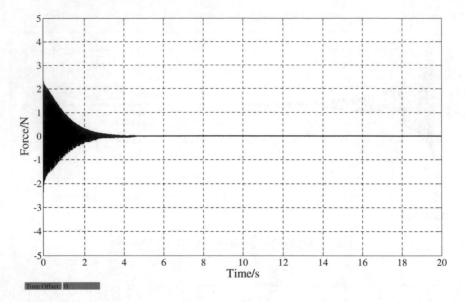

Fig. 8. Output of vibration force in time domain

constant. But after the initial coefficient was reset, the result was shown in Fig. 11, which converged to zero. It was considered that in the process of parametric convergence, it may converge to the secondary advantage, causing the degradation of vibration suppression. But in conclusion, the results illustrated that the control method with both algorithms of PID and LMS in the part of parameters setting had fulfilled the general requirement for space infrared application and the process converged quickly.

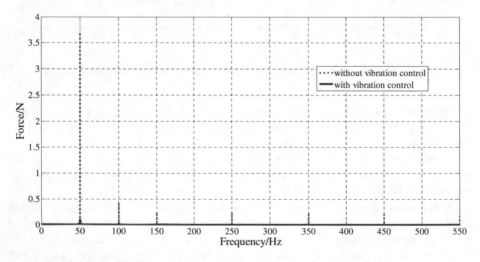

Fig. 9. The vibration spectrum of the output with and without vibration control

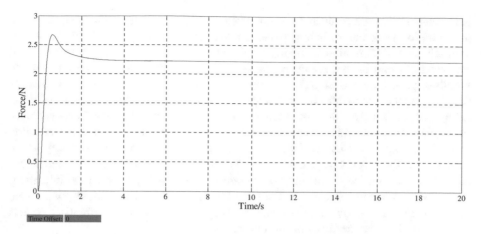

Fig. 10. Converge to the less optimum point

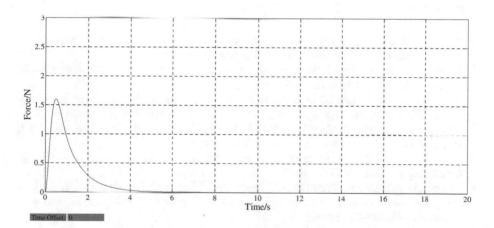

Fig. 11. Converge to the optimum point

4 Conclusion

In the space infrared remote sensors, the space mechanical cryocooler is used to ensure that infrared detectors can be in stable and low temperature conditions in the space optical remote sensing system. It is of necessity to take vibration reduction measures to meet the demand for low vibration of cryocoolers in some space application projects. In this paper a kind of control method was studied based on momentum compensation and the two algorithms of PID and LMS were given in the part of parameters setting of the control method. Finally the control method was verified by the simulation. The control method with both algorithms of PID and LMS in the part of parameters setting had fulfilled the general requirement for space infrared application and the process converged quickly. The advantages and disadvantages of the two kinds of algorithms of parameters setting are compared and analyzed as well. Simulation results showed that the vibration peak output before reduction was approximately 4.32 N while it was approximately 0.01 N after reduction by the algorithm of PID. And the first to third frequencies vibration were reduced 94%, 89% and 84%. On the other hand, the vibration peak output after reduction by the algorithm of LMS was approximately 0.005 N and the first to third frequencies vibration were reduced 96.4%, 89% and 94%. Compared the two algorithms, it is found that the algorithm of PID has the advantages of simple calculation and convenient adjustment but it is not adaptable to the variation of model parameters. However, the algorithm of LMS has a good adaptability to the model but in the process of parametric convergence, it may converge to the secondary advantage, causing the degradation of vibration suppression performance. In conclusion, in view of the inherent cryocooler model both algorithms can achieve good effect, but more in-depth study of robust and convergence of the active vibration control method will become the focus of the further research.

References

1. Pan, Y., Zhu, J.: Application and development of spaceborne cooling technology for aerospace infrared remote sensor. Philos. Vac. Cryog. **9**(1), 6–12 (2003)
2. Yang, X., Mao, N., Xu, S.: Research of cryocooler application in space infrared sensor system. Vac. Cryog. **20**(2), 113–115 (2014)
3. Yang, B., Wu, Y.: Adaptive control system for vibration harmonics of cryocooler. In: Infrared Technology and Applications XXXIX, Proceedings of SPIE, vol. 8704, pp. 1–8 (2013)
4. Riabzev, S.V., Veprik, A.M., Vilenchik, H.S., Pundak, N., Castiel, E.: Vibration-free stirling cryocooler for high definition microscopy. Cryogenics **49**, 707–713 (2009)
5. Ross, Jr. R.G.: Vibration suppression of advanced space cryocoolers—an overview. In: Proceeding of SPIE, vol. 5052, pp. 1–12 (2003)
6. Veprik, A., Riabzev, S., Vilenchik, G., Pundak, N.: Ultra-low vibration split stirling linear cryogenic cooler with a dynamically counterbalanced pneumatically driven expander. Cryogenics **45**, 117–122 (2005)
7. Doubrovsky, V., Veprik, A., Pundak, N.: Sensorless balancing of a dual-piston linear compressor of a stirling cryogenic cooler. Proc. Cryocoolers **13**(6), 231–240 (2004)
8. Wu Y.-W.A.: Stirling-cycle cryogenic cooler using feed-forward vibration control. UK Patent, 2279770 (1995)

Effect Modelling and Simulation of Earth Stray Light on Space-Born Telescope

Xuxia Zhuang$^{(\boxtimes)}$, Ningjuan Ruan, and Yan Li

Key Laboratory for Optical Remote Sensor Technology of CAST,
Beijing Institute of Space Mechanics and Electricity, Beijing 100094, China
zhuangzi111@163.com

Abstract. Space-born telescope is an important means to detect space dim objects, however, earth stray light outside FOV will reach its image plane when the camera is working, bringing the noise and creating a background. The lighted earth is a nonuniform spherical reflecting source and the amount of contamination depends on location of the sun and the satellite, sensor pointing, FOV, PST of the telescope design and so on. In order to provide reliable analysis results for bringing out the reasonable stray light mitigation scheme, quantificational modelling and simulation is needed. Firstly the characteristics of earth stray light was analysed, and base on the complex characteristics of stray light contamination, a mathematical model combined with time and space was built in this paper. Secondly, the mathematical models were translated to simulation codes, realizing effects simulation for different conditions. At last, taking a space based visible as an example, effects at different states were analysed.

Keywords: Earth stray light · Dim objects · Space based detection
Simulation analysis

1 Introduction

Space-born telescope is an important means to detect dim space objects. America, Canada and other countries developed space based visible camera to detect satellites, space debris, NEOs and so on [1–3]. With the development of technology, the demand for detection sensitivity is more urgent. Stray light is one of the most important sources affecting the sensitivity which could reduce the contrast and SNR of the image [4, 5]. The stray light could be from the sun, the moon and the earth. The sun and the moon are far away from the detection camera and could be taken as parallel source, while the earth surface is a nonuniform spherical reflecting source. The lighting condition and position of the observing platform change as time goes, causing the illuminated area by the sun and the observed area by the telescope vary. The influence on the image quality due to the earth stray varies with time and space and also there is atmosphere between the earth and the platform where the photos from the earth will be absorbed and diffused which makes this problem more complex [4, 5]. This paper aims at presenting a simulation model to analyze the effect of earth stray light on space based detection system.

© Springer Nature Switzerland AG 2018
H. P. Urbach and Q. Yu (Eds.): ISSOIA 2017, SPPHY 209, pp. 103–111, 2018.
https://doi.org/10.1007/978-3-319-96707-3_12

2   Earth Stray Light Modeling

The boresight of the space based telescope always points to the deep sky. To avoid direct sunlight, the pointing generally is anti-sun. In this condition, the reflected earth flux inevitably arrives at the camera which could contaminate the image.

The orbit of the observing platform usually is sun synchronous [2]. Take an 800 km-height orbit as an example, according to the analysis result of the beta angle, at the midwinter, the angle between the orbit plane and the sunlight is the largest. The largest sun elevation of the earth limb is about 7°, and supposing the albedo of the earth is 0.3 and the bandwidth is from 0.45 μm to 0.9 μm, the radiance at the top of the atmosphere would be 129.3 W/sr/m², which is quite serious for dim object detection system.

The earth surface is a complex extended area source which is assumed to be Lambertian. In order to calculate the image plane illumination caused by the earth stray light, the irradiance before the camera should be calculated. The illuminated area varies with the time and albedo of the earth is nonuniform, so the extended source should be subdivided to micro grids (Fig. 1).

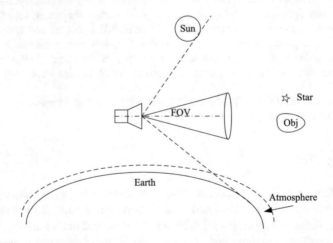

Fig. 1. Sketch of relationship between the camera and stray light source

Suppose A is the central point of the ith face of the earth, and B is the central point of the jth face of telescope aperture, and θ_{ij} is angle between normal vector of the jth face of the camera aperture and the vector from B to A, the illumination of the jth face from the ith face of the earth is,

$$E_{ij} = \frac{L_i \cdot \cos \alpha_{ij} \cdot \cos \theta_{ij} ds_i}{r_{ij}^2} \tag{1}$$

where, L_i is the radiance from the ith face of the earth depending on the earth albedo, atmosphere condition and so on, α_{ij} is the angle between the normal vector of the ith face of the earth and the vector from the point A to the point B, r_{ij} is the distance from A to B (Fig. 2).

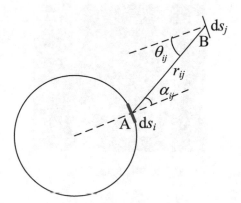

Fig. 2. Geometrical sketch

The entrance of the camera is very tiny compared with the distance from the camera to the earth, so $\theta_{i1} = \theta_{i2} = \cdots$, $\alpha_{i1} = \alpha_{i2} = \cdots$, and the illumination intensities of different points of the camera entrance from certain face of the earth are almost identical, so Eq. (1) can be approximated as follows,

$$E_i = \frac{L_i \cdot \cos \alpha_i \cdot \cos \theta_i \mathrm{d}s_i}{r_i^2} \tag{2}$$

The overall illumination intensity is integrated from the lighted and observed region of the earth, so the entrance illumination is as follows,

$$E_{entrance} = \int_{A_{vis}} E_i \mathrm{d}s_i = \int_{A_{vis}} \frac{L_i \cdot \cos \theta_i \cdot \cos \alpha_i}{r_i^2} \mathrm{d}s_i \tag{3}$$

For a given off axis angle θ_i, the Point Source Transmittance is $PST(\theta_i)$, and the illumination arrives at the image plane can be written as follows,

$$E_{image} = \int_{A_{vis}} E_i \mathrm{d}s_i = \int_{A_{vis}} \frac{L_i \cdot \cos \theta_i \cdot \cos \alpha_i}{r_i^2} PST(\theta_i) \mathrm{d}s_i \tag{4}$$

Where, A_{vis} is the region of the earth being lighted and projected from the camera. The earth is divided into many micro grids, and each grid has its normal vector. The normal vector of the ith grid is $\vec{e_i}$, and the unit vector of the sunlight is $\vec{s}$. If the sun elevation is larger than 0, we can say this area is lighted, so the illuminated region is,

$$A_{sun} = \{E, \overrightarrow{s} \cdot \overrightarrow{e_i} > 0\} \qquad (5)$$

Where, E is the surface of the earth, and A_{sun} is the lighted area of the earth (Fig. 3).

Fig. 3. The lighted region

The earth area projected from the camera is shown in Fig. 4, and can be written as $A_{sat} = \{E, \left(\overrightarrow{se_i} \cdot \overrightarrow{p} \geq 0\right) \cap (r_i \leq R_{max})\}$, and $A_{vis} = A_{sun} \cap A_{sat}$, which is the lighted area projected from the camera. Where, E is the earth surface, $\overrightarrow{se_i}$ is the vector from central point of the ith face of the earth to the satellite, $\overrightarrow{p}$ is the vector of boresight of the camera, r_i is the distance from the earth to the satellite, and R_{max} is the distance from the satellite to the earth limb point.

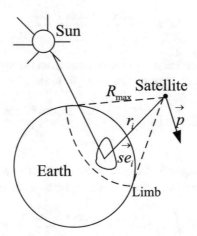

Fig. 4. The lighted area projected from the camera

θ, α_i and r_i vary with time, orbit and light of sight, and is time-variant and space-variant. This paper will use the satellite orbit analysis tool to calculate satellite positions, sun elevations and so on. The radiance from the earth is depend on earth albedo

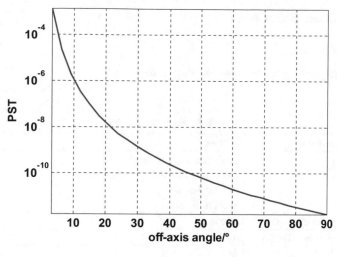

Fig. 5. PST

distribution, sun elevations, observing angles, conditions of atmosphere and so on, and to cope with this question, the MODTRAN will be used to calculate the radiance matrice at the top of the atmosphere.

3   Simulation of the Earth Stray Light

The simulation procedure of the earth stray light is as follows,

Initialize the imaging time, parameters of satellite, sensor boresight, field of view and so on;

(1)   Calculate the earth region lighted by the sun and the corresponding sun elevations;
(2)   Calculate the angle matrice between the normal vectors of the earth and the vectors from the earth grids to the satellite, the angle matrice between the bore-sight and vectors from the satellite to the earth grids and the distances from the satellite to the earth grids.
(3)   Calculate the visible area by the camera, and find the intersection of the area illuminated and the observable region;
(4)   Import the global albedo;
(5)   Calculation the radiance matrice reaching at the top of the atmosphere from the illuminated earth surface;
(6)   Call the Eq. (3) to calculate the illumination at the aperture;
(7)   Call the Eq. (4) to calculate the illumination at the image plane.

4 Simulation Example

1. Basic parameters

The basic parameters are listed in the Table 1.

Table 1. Basic parameters

Parameters	Values
Orbit	Sun-synchronous orbit 18:00 local time descending node 98° inclination 800 km altitude
Aperture diameter	200 mm
Spectral bandwidth	0.45–0.9 μm
Field of view	6°
Elevation of boresight	−10° (up), 0° (horizontal), 10° (down), 15° (down)
Atmosphere	1976 US Stardard Rural-VIS = 23 km

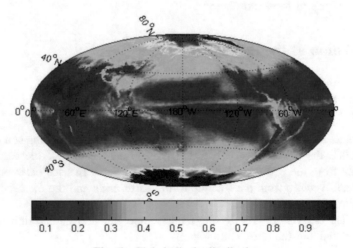

Fig. 6. Global albedo distribution

The PST is as follows,

The global albedo uses reflectivity data of a remote sensing satellite called TOMS [6], which is shown in Fig. 6.

2. Simulation results

The sun elevation of the earth limb at different season is shown in Fig. 7.

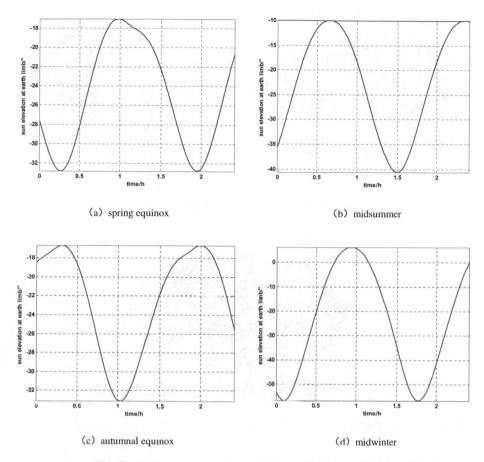

(a) spring equinox

(b) midsummer

(c) autumnal equinox

(d) midwinter

Fig. 7. The sun elevation the earth limb at different seasons

Figure 7 shows that in the midwinter, the sun elevation at the earth limb is the largest, which means the earth stray light is the most serious. So the earth stray light at the midwinter will be emphatically analyzed.

Figure 8 gives the sun elevation distribution for different positions of the earth at certain time in midwinter. The dark area is the area visible to the camera. The second figure shows the edge of the area. In the figures, the sun elevations below zero are set to zero. From the result, we can see that the lighted area visible to the camera is near the South Pole, where is covered by ice and snow with high albedo.

The radiance at top of the atmosphere from the lighted area is shown in Fig. 9.

Based on results of the radiance at the top of the atmosphere, observation elevations, PST and so on, the illumination at the aperture and image plane is shown in Table 2.

If PST rises to be ten times of the prior shown in Fig. 5, the illumination at the aperture and image plane is shown in Table 3.

The results show that illumination at the image plane rises with the PST.

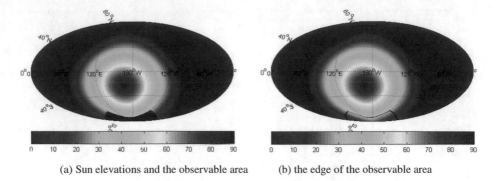

(a) Sun elevations and the observable area   (b) the edge of the observable area

Fig. 8. The sun elevation of the earth at the midwinter

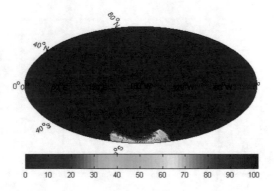

Fig. 9. Radiance distribution at the top of the atmosphere from the observable area (W/sr/m^2)

Table 2. Illumination intensity at the aperture and image plane for different pointing

Elevation of boresight (°)	At the aperture (W/m^2)	At the image plane (W/m^2)
−10°	30.6	1.1×10^{-9}
0°	45.6	4.7×10^{-9}
10°	63.4	3.3×10^{-8}
15°	73.2	1.3×10^{-7}

Table 3. Illumination intensity at the aperture and image plane for different pointing

Elevation of boresight (°)	At the aperture (W/m^2)	At the image plane (W/m^2)
−10°	30.6	1.1×10^{-8}
0°	45.6	4.7×10^{-8}
10°	63.4	3.3×10^{-7}
15°	73.2	1.3×10^{-6}

From the above results, we can conclude that,

There are several factors affecting the earth stray light such as imaging time, elevation of boresight and PST of the camera. In the midwinter, the stray light is the most serious, when the illumination intensity at the aperture would be tens of W/m^2. PST of the observing camera is also one of the most important factors affecting the earth stray flux reaching to the image plane of the telescope. In order to reduce the amount of the earth stray illumination, the PST should be as low as possible. And also the pointing of the camera should be far away from the earth to reduce the earth stray light.

5 Conclusion

A simulation model to analyze the effect on image quality of the space born telescope was built in this paper, which could be used to do effect simulation at different time, positions, orbits, pointing, PSTs for different telescopes. The simulation results could be used to predict the image quality reduced by earth stray light, and give proper requirement for stray light rejection.

References

1. Lambour, R., Bergemann, R., et al.: Space-based visible space object photometry: initial results. J. Guid. Control Dyn. **23**(1), 159–164 (2000)
2. Hackett, J., Brisby, R., Smith, K.: Overview of the sapphire payload for space situational awareness. Proc. SPIE **8385**, 1–10 (2012)
3. Laurin, D., Hildebrand, A., Cardinal, R., et al.: NEOSSat - a canadian small space telescope for near earth asteroid detection. Proc. SPIE **7010**, 1–12 (2008)
4. Chunming, Z., Yongchun, X., Li, W., et al.: Analysis of influence of earth albedo on star tracker. Laser Infrared **42**(9), 1011–1015 (2012)
5. Xiangguo, X., Zhonghou, W., Jiaguang, B., et al.: Influence of earth radiation on photoelectric detection system based on space. Acta Photonica Sin. **38**(2), 375–381 (2009)
6. Dan Bhanderi, D.V.: Modeling earth albedo for satellites in earth orbit. In: AIAA Guidance, Navigation, and Control Conference And Exhibit, California (2006)

Thermal Optical Analyses on a Space Camera Based upon the Vacuum Test

Z. M. Zhao$^{(\boxtimes)}$, B. Yu, and Y. Su

Beijing Institute of Space Mechanics and Electricity,
Key Laboratory for Advanced Optical Remote Sensing Technology of Beijing,
Beijing 100094, China
13693205256@163.com

Abstract. Based upon the temperature data from the thermal balance test in vacuum environment, thermal optical analyses on a space optical remote sensor (space camera), aiming at investigating the thermal influences on the optical performance, were carried out. In this paper, the measured discrete temperature data were applied to obtain the integrated temperature profile of the camera, and the thermal distortions and optical performances with that temperature distributions were calculated, which was different from the general analyses procedure published in the open literatures. By the researches of inversion temperature field analyzing, thermal distortion analyzing, optical performance analyzing, and data comparing between simulation and test, it was indicated that temperature is one of the effect factors to the optical performance of this space camera. The results also indicated that the thermal distortion influences on the optical performance could be farthest restrained by controlling the temperature of the main structural and optical components under the design value.

Keywords: Space camera · Thermal balance test · Vacuum test
Thermal optical analysis

1 Introduction

With the evolution of earth observation and universe exploration, the performances and image qualities of space camera are improved as well. Therefore, high resolution (HR) optical instruments are worked out and become the main payload of the remote sensing spacecraft during these decades. In order to obtain the high resolution images, accurate optical/structural/thermal designs for the space camera are demanded. Previous researches indicated that temperature was one of the significant factors for the space camera, and even slight changes in temperature could give rise to thermal deformations, which were further translated into imaging quality variation. Consequently, the thermal optical analyses, also known as the structural-thermal-optical performance (STOP) analyses [1, 2], became the important method in predicting the effect of thermal distortion on optical performance. During the past decades, a number of open literatures about the thermal optical and STOP analysis on the space camera were published, which could be in references [3–6].

© Springer Nature Switzerland AG 2018
H. P. Urbach and Q. Yu (Eds.): ISSOIA 2017, SPPHY 209, pp. 112–122, 2018.
https://doi.org/10.1007/978-3-319-96707-3_13

In this paper, the thermal optical analyses on a space camera based upon the thermal balance and imaging test in vacuum environment were carried out. The space camera was a three-mirror, visible light camera under development by a team consisting of Beijing Institute of Space Mechanics & Electricity, directly under the China Academy of Space Technology (CAST). The camera, as shown in Fig. 1, was composed by three main elements: (1) the front optical components including the primary mirror (M1) and the secondary mirror (M2); (2) the tertiary mirror (M3) components; (3) the focal plane unit (FPU) consisting of visible charge couple device (CCD) and the electrical circuit.

In order to insure the high optical quality, detailed thermal design was performed on this camera by using multilayer insulations, heat pipes, thermal surface finishes, heaters etc.

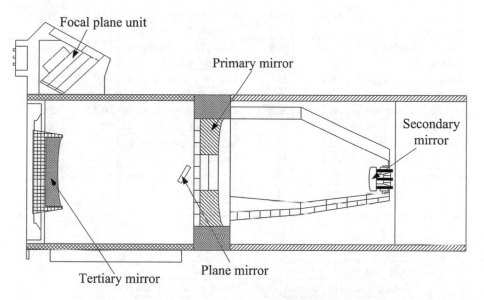

Fig. 1. Schematic drawing of the space camera

2 Experimental Apparatus and Measurement Procedures

In thermal balance test, there was a highly complex test facility designed to satisfy the test objectives and to provide the most approximate space simulation. Therefore, the space environment (vacuum, 4 K space heat sink and the orbit heat flux) simulation test system and the high accuracy data acquisition system were built to obtain the test environment and to measure the temperature field and optical performance of the camera. The temperature measurement device was custom-built, which could both measure the temperature signals from the thermostats and control the heating power of the heaters based upon the measured temperature.

Two test cases were carried out, including the worst hot case and the worst cold case. Table 1 is the measured temperatures of the camera and Table 2 is the optical performances under the two test cases. In Table 2, the positive value of focal plane location means the focal plane moved towards to the M1 while the negative value means backwards movement to M1 in tests.

Table 1. Measured temperatures of the camera

Location	Cold case (°C)	Hot case (°C)	Location	Cold case (°C)	Hot case (°C)
M1 baffle #1	16.1	20	Second mirror #2	18.3	19.8
Second mirror #1	20	20.2	M1 hub #2	20.3	20.5
Main support structure	20.1	20.1	M1 hub #3	20	20.3
M1 hub #1	20	20.2	M1	19.3	19.9
Second mirror support structure #1	18.8	20.4	Second mirror support structure #4	16.9	19.6
Second mirror support structure #2	19.4	20.9	Second mirror support structure #5	16.6	19.6
Second mirror support structure #3	19.3	21	Second mirror support structure #6	16.9	19.7
M3 support structure #1	20.2	20.1	M3 support structure #3	20.5	20.4
M3 support structure #2	19.9	20	M3 support structure #4	20	20.1
Tertiary mirror #1	20.2	20.2	M3 support structure #5	18.9	18.9
M1 baffle #2	16.2	20	Camera support frame	25.4	25.8
M1 baffle #3	13.2	19.8	FPU support structure	17.2	17.5
M1 baffle support structure	17.7	21.2	Tertiary mirror #2	20	20

Table 2. Measured optical performances

Cold case		Hot case	
Focal plane location (mm)	MTF	Focal plane location (mm)	MTF
−0.13	0.155	−0.53	0.159
−0.08	0.174	−0.48	0.172
−0.03	0.185	−0.43	0.181
0	0.19	−0.41	0.182
0.025	0.1875	−0.38	0.178
0.07	0.174	−0.33	0.167
0.12	0.161	−0.28	0.148

3 Thermal Optical Analyses

Figure 2 showed the procedure of the thermal optical analyses based upon the vacuum test. Unlike the general thermal optical analyses procedure published in the open literatures, the current paper used the discrete spots of temperature measured in vacuum test to calculate the entire temperature distributions of the camera. Through the inversion research, the real thermal status of the camera could be obtained for the thermal distortion simulation, and then the optical analyses. The thermal optical analyses procedure in this paper mainly includes four key steps:

- Inversion research on temperature field under the measured discrete temperatures, the heating power of the heaters on camera, the test environment and the orbital heat flux
- Thermal distortion analyses under the calculated temperature field
- Zernike polynomial fitting and optical analyses
- Data comparison and thermal optical analyses model modification.

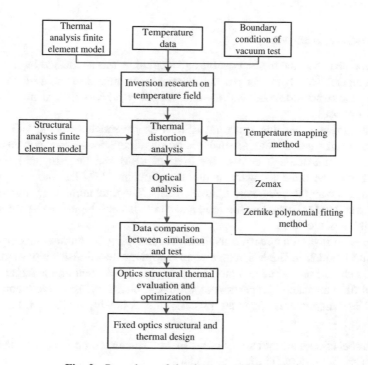

Fig. 2. Procedure of the thermal optical analysis

3.1   Thermal Analyses (Inversion Research on Temperature Field [7])

As mentioned above, the inversion thermal analyses used measured discrete temperatures, the heating power of heaters, the test environment and the orbital heat flux as the boundary conditions and initial conditions to calculate the real temperature field of the camera in the thermal vacuum test. The mathematical models of the inversion thermal analyses included an I-DEAS TMG finite element model (FEM) for temperature field calculation and an I-DEAS TMG script code for data comparing and convergence control automatically.

The geometry of thermal FEM model was consisted of several components, including the space camera, the satellite module and the IR array test facility, as shown in Fig. 3. In the thermal FEM model, the shell elements were used to simulate all the support structures, and the solid elements were created for all the mirrors with shell elements covering on its entire surfaces. The total number of the elements was 7894.

The I-DEAS TMG script code was a secondary development on the I-DEAS commercial software written by FORTRAN computer language. For each iteration loop, the script code compared the numerical temperature results with the test data automatically and adjusted the heating power of each heater for the next loop. The solution was converged whenever all the differences between the numerical temperature of the sensor element and the measured temperature of the thermistor were less than convergence error.

3.2   Structural Analyses

In structural analyses, the thermal distortion calculation was conducted by mapping the camera temperatures obtained from the thermal math model to the structural model. Therefore, the structural model and the data mapping process were demanded for the structural analyses.

In the current paper, the structural FEM was generated with the UG NX commercial software and all the thermal distortion simulations were solved with the UG NX Nastran. Although both of the thermal FEM and the structural FEM were generated from the same geometry model, the structural FEM had more details of the camera structure and thereby more finite elements. The total number of structural FEM elements was 210,000 including the solid and shell element. Figure 4 was the structural FEM of the space camera.

There were several approaches available for mapping the thermal model results to the structural model. In UG NX, temperature mapping created associations between the element's centroid on the thermal model and the closest element on the target model. If the nodes did not match, temperatures were interpolated by using the elements' CG. In UG NX temperatures mapping process, the following general rules must be followed:

- The global coordinate systems of target FEM (structural model) must be the same as that in the source model (thermal model).
- Both models should be geometrically congruent but did not need to have the same mesh.
- Mapped temperatures were written in a certain format defined by UG NX.

3.3   Data Exchange Using Zernike Polynomials and Optical Analyses

The last step of thermal optical analyses procedure was the optical performance calculation and evaluation. Zernike polynomials was used to transfer the optical surface deformations data to the optical numerical code. There are three advantages for this method: (1) Zernike terms which was directly related to Seidel aberrations was an effective way to estimate the imagery quality of the optical system; (2) Zernike polynomials had some very helpful mathematical properties, such as sole, orthogonal and independent, which mad them a competent choice to describe wave front error of optical elements; (3) Zernike polynomials could describe the distortion surface quite accurate [8].

Though the Zernike polynomials could have as many items as you wanted to obtain higher fitting accuracy of the distorted camera mirror surfaces, the current paper chose the first 28 Zernike coefficients which would be excellent enough to fit every attempted surface.

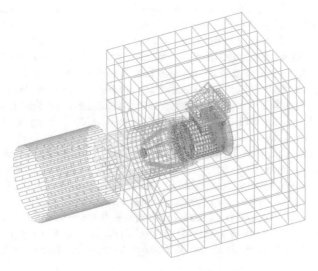

Fig. 3. Thermal analysis finite element model generated by I-DEAS

Fig. 4. The structural finite element model of the space camera

In these analyses, the Zernike polynomials were calculated by an in-house code which could automatically transfer the Zernike coefficients to the Zemax commercial software for optical performance calculating.

4   Results and Discussions

In the thermal balance test, the hot case was conducted primarily. When the camera temperatures meet the test requirements, all the temperatures measured by the thermistors and the optical performance parameters including focus location and Modulation Transfer Function (MTF) were recorded. In this case, the optimal focus location with maximum MTF was set to the reference focus location, and all the other measured focus locations were the relative data compared with the reference focus location. After the data recording in hot case, the cold case tests which duplicated the same procedure as hot case started, and all the focus locations were the relative data compared with the reference focus location as well. For better comparison between the simulation results and test data, the numerical work chose the same data processing approach of focus location as the imagery test.

Figures 5 and 6 present the cut-away view of inversion temperature profile for the cold case and hot case, respectively. As shown in these figures, in cold case, all the temperatures of space camera except the M2 support structure were within 20 °C ± 2 °C and the temperatures of M2 support structure were less than 17 °C. However, in hot case, the temperatures of the entire camera components were within 20 °C ± 2 °C. This temperature variation leads to a serious thermal distortion on M2 support structure, as shown in Fig. 7 and then became one of the significant effect factors to the optics.

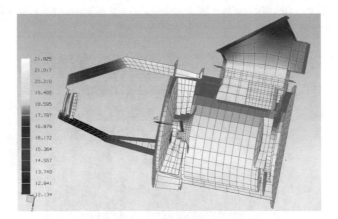

Fig. 5. Temperature profiles, cold case

As the input of Zernike polynomial calculation, the absolute coordinates and displacements of each node on mirror surfaces were required. As the example of data processing method, Fig. 8 presents the displacements of mirror surfaces in cold case and Table 3 shows the data format of node coordinates and displacements.

Fig. 6. Temperature profiles, hot case

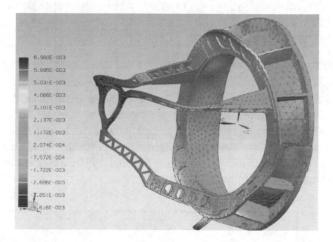

Fig. 7. Thermal distortion of the main support structure and the secondary mirror support structure, cold case (magnification ratio: 1000x)

Fig. 8. Displacements of mirror surfaces, cold case

Table 3. Node displacements in cold case (sectional)

Location	Node ID	X	Y	Z	ΔX	ΔY	ΔZ	Amplitude
Primary mirror	634935	13.1	134.1	75.4	8.44E−04	4.45E−05	−7.35E−05	8.48E−04
	634936	10.4	121.4	65.8	8.31E−04	4.09E−05	−6.94E−05	8.35E−04
	634937	8.1	109.3	56.2	8.19E−04	3.78E−05	−6.67E−05	8.23E−04
	…	…	…	…	…	…	…	…
Secondary mirror	634790	349.2	3.6	−38.3	−2.95E−03	−1.45E−04	3.33E−03	4.45E−03
	634791	348.3	6.9	−30.9	−3.09E−03	−1.66E−04	3.32E−03	4.54E 03
	634792	347.7	10.1	−23.4	−3.23E−03	−1.87E−04	3.31E−03	4.63E−03
	…	…	…	…	…	…	…	…
Tertiary mirror	624934	−315.1	−35.7	71.8	2.41E−04	−6.98E−05	7.56E−05	2.62E−04
	624935	−316.8	−26.4	65.9	2.60E−04	−6.53E−05	7.53E−05	2.79E−04
	624938	−320.3	0.7	47.9	3.16E−04	−5.44E−05	7.70E−05	3.30E−04
	…	…	…	…	…	…	…	…
Plane mirror	635387	−30.8	15.3	−15.8	−3.95E−04	−3.90E−04	3.32E−04	6.46E−04
	635390	−36.4	15.3	0.7	3.39E−05	−4.29E−04	−2.89E−05	4.32E−04
	635391	−38.3	15.3	6.3	1.41E−04	−4.54E−04	−1.53E−04	5.00E−04
	…	…	…	…	…	…	…	…

Figures 9 and 10 presented the MTF versus focus location in cold and hot case, respectively. In these figures, the design parameters, the thermal optical simulation results and the test data were presented for comparison. It can be concluded from Fig. 9 that the thermal optical simulation results agreed well with the design parameters and test data, which could prove the validity of the thermal optical analysis models and data processing method applied in the current paper.

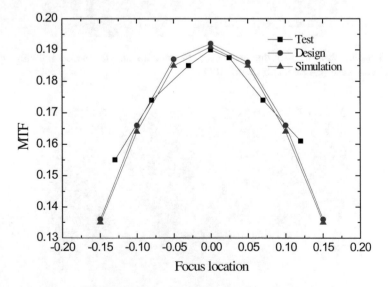

Fig. 9. MTF versus focus location, cold case

There was a great deviation between the test data and the design parameters in Fig. 10, This can be attributed to temperature variation from the design values which can induce a significant thermal distortion on structures and mirrors. The thermal optical analyses results represent this phenomenon as well.

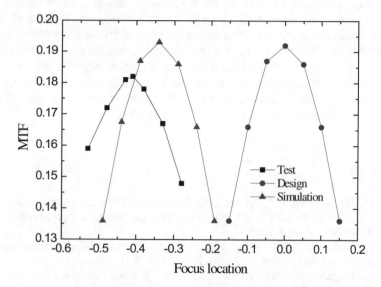

Fig. 10. MTF versus focus location, hot case

Table 4 summarized the measured and calculated optimal focus and corresponding MTF in cold and hot case.

Table 4. Optimal focus and MTF comparison between the simulation results and test data

Case	Optimal focus (mm)		MTF	
	Test	Simulation	Test	Simulation
Cold case	0	0.082	0.190	0.192
Hot case	−0.41	−0.3449	0.182	0.193

5 Conclusions

Thermal optical analyses on a space camera based upon the vacuum test had been carried out to investigate the thermal influence on the optical performance and all the analyses process and simulated results had been described in this paper. From the analyses results, major findings of this study were as follows:

1. From the data comparison among the thermal optical analysis results, the optics design parameters and the measured optical performances in vacuum test, it could

be concluded that the analyses approaches, including the inversion research on temperature field, the thermal distortion calculation, the Zernike polynomials fitting and the optical analyses, conducted in the current paper can predict the thermal optical analysis accurately.

2. The thermal distortion caused by the temperature variation could influence space camera optical performance significantly. Therefore, the exact thermal design must be conducted on camera to control the temperature of the main structural and optical components under design value for farthest restraining the thermal distortion.

3. For this camera, the thermal distortions of the main support structure and the secondary mirror support structure are the significant effect factor of the optical performance. Higher precision thermal control method should be adopted to avoid optical performance exacerbated arose by defocusing.

References

1. Johnston, J.D., Howard, J.M., Mosier, G.E., Parrish, K.A., Mcginnis, M.A., et al.: Integrated modeling activities for the James Webb space telescope: structural-thermal-optical analysis. Proc. SPIE **5487**, 600–610 (2004)
2. Kunt, C., Mule, P.: WFC3 optical bench structural thermal optical performance (STOP) analysis and optimization. In: AIAA, pp. 2003–1530 2003
3. Cho, M., Corredor, A., Vogiatzis, K., Angeli, G.: Thermal performance prediction of the TMT optics. Proc. SPIE **7017**, 701716 (2008)
4. Chen, Z.P., Chen, Z.Y., Yang, S.M., Shi, H.L.: Thermal calculation and structure analysis of space main optical telescope. Proc. SPIE **6687**, 66870Y (2007)
5. Zurmehly, G.E., Hookman, R.A.: Thermal and structural analysis of the GOES scan mirror's on orbit performance. SPIE Anal. Opt. Struct. **1532**, 170–176 (1991)
6. Weinswig, S., Hookman, R.A.: Optical analysis of thermal induced structural distortions. SPIE Curr. Dev. Opt. Des. Opt. Eng. **1527**, 118–125 (1991)
7. Zhang, X.C., Shuai, Y., Qing, H.X., Tan, H.P.: Inversion research on temperature field with nonlinear multiple heat source using I-DEAS. J. Astronaut. **9**, 2088–2095 (2009)
8. Li, J.H., Wang, J.Q., Lu, E., Wu, Q.W.: Thermal-optical analysis and thermal design of a space telescope. In: AIAA, pp. 2000–0736 (2000)

Structural Design and Analysis of Freeform Surfaces in Off-Axis Reflective Space Optical Imaging Systems

Xiaoxiao Zhang$^{(\boxtimes)}$, Shengbang Yuan, and Chunnan Wu

Beijing Institute of Space Mechanics and Electricity,
Engineering Technology Research Center for Aerial Intelligent Remote Sensing
Equipment of Beijing, Beijing 100094, China
Cathy_xiaoxiao@126.com

Abstract. Freeform surface has fully with the design ability of off-axis reflective imaging system, and has the potential to achieve higher performance indicators, the complex structural constraints, meeting the various high or special needs in the field of space optics. Either at home and abroad, related research groups have been successfully designed or developed many free surface off-axis reflective imaging system. They have advance performance indicators, or have the special system structure, adhere to the two application directions in the field of space optical system of freeform surface, to improve the optical performance and to realize the special structure. The article particularly introduce the structural design and analysis of freeform surfaces in off-axis reflective space optical imaging system.

Keywords: Freeform surface · Off-axis reflective
Structural design and analysis

1 Introduction

In recent years, in the field of space optics closely related to the aerospace engineering, off-axis reflective imaging system has received more and more attention, due to its wide spectrum SNR, high S/N ratio, large distance of target detection and recognition, flexible of material selection [1]. However, the rotational symmetry of off-axis reflective imaging system naturally has a series of asymmetric aberration, and the aberration cannot usually be corrected by the traditional spherical or aspherical design [2–5]. Therefore, using spherical or aspherical method to design the off-axis reflective system is very difficult, especially for the space optical system with high performance requirements and complex structure constraints.

Freeform surface is a kind of rotational symmetry of non-traditional optical surface, which is a research hotspot in the field of current international optical design. The design can provide optical designers with higher degrees of freedom, helping to achieve the design of high performance system in excellent image quality, advanced indicators, compact structure, and small size. At the same time, as a kind of non rotational

© Springer Nature Switzerland AG 2018
H. P. Urbach and Q. Yu (Eds.): ISSOIA 2017, SPPHY 209, pp. 123–133, 2018.
https://doi.org/10.1007/978-3-319-96707-3_14

symmetry surface, freeform surface also has the ability of correcting asymmetric aberration of off-axis system [6, 7].

2   Profile of Optical Free Surface and Its Application in Space Imaging System

Optical freeform surface refers to the rotational symmetry of unconventional optical surface, that do not have the global rotational symmetry, no unified optical axis, usually cannot be described by the spherical or aspherical coefficient. It can be broadly included in the following categories:

(1)   no symmetry axis complex unconventional continuous curved surface;
(2)   discontinuous surface shape mutation surface;
(3)   the steradian big curved surface.

The first kind of Freeform surface in the field of space optical imaging has important application value. This kind of Freeform surface including bracelet surface complex surface, XY polynomial surface Zernike polynomial surface non - uniform rational b - splines, etc. At present, the domestic and foreign existing design or physical systems mostly adopt this kind of free surface, the continuous form can be more easily to realize in processing, detection and adjustment, especially the XY polynomial surface and Zernike polynomial surface.

In the aspect of space optical imaging system, free-form surface technology can be mainly applied in the following respects:

(1)   Off-axis reflective imaging system of small F number;
(2)   Off-axis reflective imaging system of large view field angle;
(3)   Off-axis reflective imaging system of with an exit pupil;
(4)   Afocal off-axis reflective imaging system;
(5)   Off-axis reflective imaging system of compact structure.

3   Structural Design of Freeform Surface Off-Axis Reflective Optical Imaging System

Traditional coaxial reflective optical systems usually have viewing angles below 3°, and with bigger F numbers. To the contrary, freeform surface off-axis three-mirror optical systems are able to meet the demand of viewing angle and small F number. In addition, the structure of off-axis three-mirror optical system structure has the characteristics of asymmetric, so the layout is very flexible. This scheme selects the freeform surface off-axis three-mirror optical system, the overall configuration is shown in Fig. 1. Light goes into the optical window and 45° reflector, then into the off-axis three-mirror optical system. Three-mirror system uses wide field on sagittal direction, prone on the top of the bucket or ball.

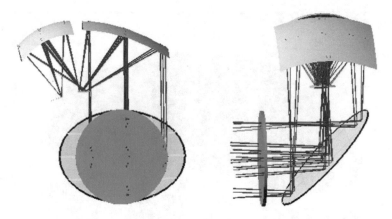

Fig. 1. Overall configuration of optical system

The primary mirror (M1) and third one (M3) of system are free-form surfaces. The mirror surfaces are in the form of 6 order XY polynomial surface. Because the system is meridian plane of symmetry, so odd coefficient of x is 0. The surface mathematical expression is as follows

$$
\begin{aligned}
z(x,y) = {} & \frac{c(x^2+y^2)}{1+\sqrt{1-(1+k)c^2(x^2+y^2)}} + A_2 y + A_3 x^2 + A_5 y^2 + A_7 x^2 y \\
& + A_9 y^3 + A_{10} x^4 + A_{12} x^2 y^2 + A_{14} y^4 + A_{16} x^4 y + A_{18} x^2 y^3 + A_{20} y^5 \\
& + A_{21} x^6 + A_{23} x^4 y^2 + A_{25} x^2 y^4 + A_{27} y^6.
\end{aligned}
\tag{1}
$$

Z for surface rise, c as the surface curvature, k for quadric coefficient, A is the coefficient of polynomial of the i th.

Secondary mirror is of 8 order off-axis aspheric surface mirror. Its mathematical expression is as follows

$$
z(x,y) = \frac{c(x^2+y^2)}{1+\sqrt{1-(1+k)c^2(x^2+y^2)}} + A(x^2+y^2)^2 + B(x^2+y^2)^3 + C(x^2+y^2)^4
\tag{2}
$$

3.1 The Overall System Solution of Free-Form Surface Optical and Mechanical Structural Design

According to the results of the optical system design, system working conditions, the optical system installation requirements within the pod, optical system and environment conditions inside the pod and other various constraints, the design of the free surface optical and mechanical structural design of the scheme is shown in Fig. 2.

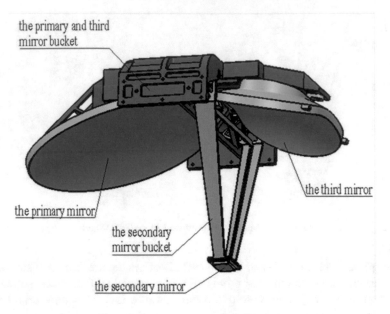

Fig. 2. Design of the optical and mechanical structure

Optical system as an independent component, is installed in the system within the pod on bearing axis framework. Minimizing the other components in the design of its stability, to ensure that the optical system imaging quality, and be more convenient within the pods of the loading test. As the need of the optical system for the spectral radius of high reflectivity, in order to increase the optical transmittance, reduce the energy loss effectively, three mirror reflectors are high reflective gold plating membrane, reflectivity of 98%. In order to improve the imaging quality of optical system, we use integrated design of the optical system, structure, material, thermal environment, and mechanical environment, to simplify structure and avoid repetition structure, with the smallest weight to gain maximum strength and stiffness.

Considering the major advantages of the aluminum material in high thermal conductivity, better thermal diffusion performance, mature processing and coating process, shorter processing cycle, non-toxic harmless, lower cost, as the reflector material can realize the integration of mirror and frame design, simplify the optical adjustment process, we select aluminum material as the basis of optical reflector. But its shortcomings in big density, larger density, lower stiffness and bigger thermal expansion coefficient, therefore, using aluminum as a mirror material, mirror weight reduction design, heat dissipation design and thermal control design of the system must be considered.

3.2   The Structural Design of Lens Body

According to the shape and size of the mirror, vertex curvature radius, effective aperture, which have been determined in the scheme design of optical system, optical caliber reflectors are designed. In the primary mirror, for example, the design scheme is shown in Fig. 3. Primary mirror is installed in the way of the back of the lens body by three points, bolted on the main three mirror bracket, using pin positioning adjustment in place. At the same time, three unload groove is designed installation support position, ensure the external stress make smaller influence on mirror decent shape.

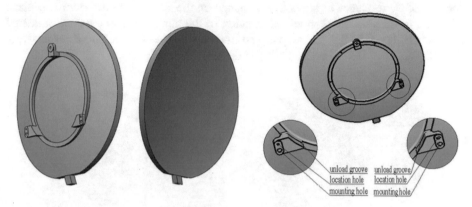

Fig. 3. Design of the primary mirror

3.3   The Design of the Support Structure

According to the results of optical design, selecting rational mirror component structure so as to ensure the imaging quality of optical system. In the process of reflector component structural design, minimizing the influence of environmental temperature changes on the system optical quality, in guarantee under the premise of optical system and mechanical performance requirements, to make lightweight and integration design, at the same time to consider processing manufacturability, assembly and debugging and testing of convenience.

The primary and third mirror bracket structure as shown in Fig. 4(a), primary and third mirror bracket is connected to the bearing axis framework of pod, taking side with screw. Positive on both ends of each design has three coplanar convex platform used for installation of the primary mirror and third mirror. The secondary mirror bracket structure as shown in Fig. 4(b).

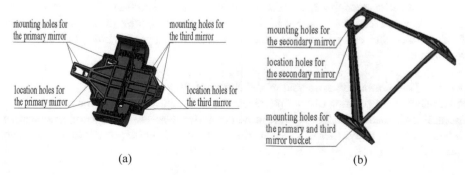

(a)   (b)

Fig. 4. Design of the primary and third mirror bucket, design of the secondary mirror bucket

4   Structural Analysis of Freeform Surface Off-Axis Reflective Optical Imaging System

4.1   The Simulation Analysis of Key Parts

4.1.1   The Simulation Analysis of the Lens Body

Because of the influence of temperature change on the aluminum mirror surface shape is bigger, to analyzes the thermal deformation of the mirror, due to the environmental temperature is 20 ± 0.5 °C in pod, so thermal analysis simulation respectively $20 + 0.5$ °C, $20 - 0.5$ °C deformation of the two states (Fig. 5).

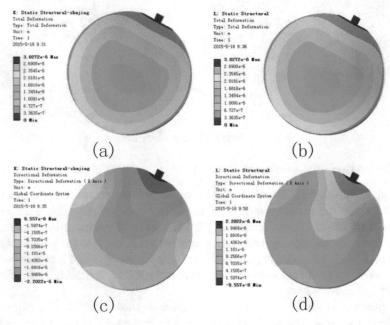

Fig. 5. Thermal deformation of the primary mirror. (a) $20 + 0.5$ °C; (b) $20 - 0.5$ °C; (c) $20 + 0.5$ °C perpendicular to the mounting surface; (d) $20 - 0.5$ °C perpendicular to the mounting surface.

Thus it can be seen that the primary mirror has large distortion of the lens, near the 2 μm PV value requirements, but in the direction perpendicular to the mounting surface, causes the curved surface shape of small changes, within the range of requirements.

4.1.2   The Simulation Analysis of the Support Structure

As a result of the position of the optical system with adjustable precision is higher, the thermal deformation of the support structure will cause the change of the relative location of lens body, thermal deformation analysis of support structure on the two states of $20 + 0.5$ °C, $20 - 0.5$ °C have be done (Fig. 6).

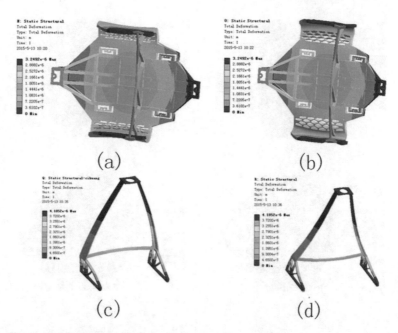

(a)   (b)

(c)   (d)

Fig. 6. Thermal deformation of the bracket. (a) 20 + 0.5 °C of the primary and third mirror bracket; (b) 20 − 0.5 °C of the primary and third mirror bracket; (c) 20 + 0.5 °C of the secondary mirror bracket; (d) 20 − 0.5 °C of the secondary mirror bracket.

Thus it can be seen that the deformations of the primary and third mirror bracket, and the secondary mirror bracket are all within the tolerance, that can meet the requirements (Fig. 7).

The modal analysis of the primary and third mirror bracket, and the secondary mirror bracket are shown in Figs. 8 and 9.

Thus it can be seen that the the primary and third mirror bracket of 1 order natural frequency is 218.04 Hz, secondary mirror bracket of 1 order natural frequency is 125.97 Hz, all meet the requirements of 1 order natural frequency greater than 80 Hz, the indicators meet the design requirements.

4.2 Analysis of Optical System Design and Performance

4.2.1 The Thermal Analysis of Optical System

The optical system in the 20 ± 0.5 °C thermal deformation as shown in Fig. 10. The maximal displacement of the optical system can be seen that is 2.58 μm. Larger deformation on the edge of the lens body is mainly due to the causes of the lens body material, but in the direction perpendicular to the underside of deformation is small, only 1.54 μm, in the range of 2 μm, can meet the requirements.

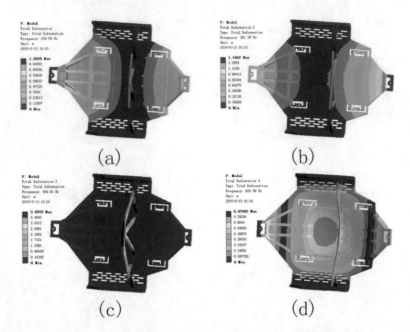

Fig. 7. Modal analysis of the primary and third mirror bracket. (a) Shape of the 1 order modal; (b) Shape of the 2 order modal; (c) Shape of the 3 order modal; (d) Shape of the 4 order modal.

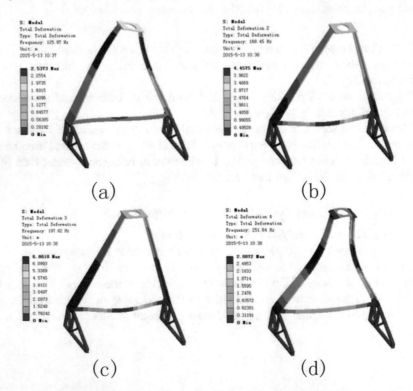

Fig. 8. Modal analysis of the secondary mirror bracket. (a) Shape of the 1 order modal; (b) Shape of the 2 order modal; (c) Shape of the 3 order modal; (d) Shape of the 4 order modal.

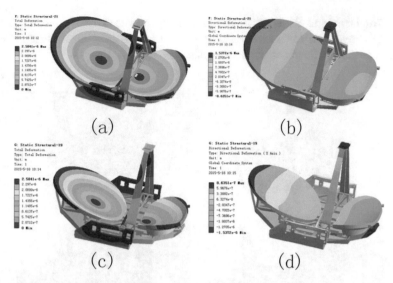

Fig. 9. Thermal deformation of the optical system. (a) 20 + 0.5 °C; (b) 20 − 0.5 °C; (c) 20 + 0.5 °C perpendicular to the mounting surface; (d) 20 − 0.5 °C perpendicular to the mounting surface.

4.2.2　The Modal Analysis of Optical System

Therefore, the optical system of 1 order natural frequency is 110.59 Hz, that meet the requirements of 1 order natural frequency greater than 80 Hz, the indicators meet the design requirements.

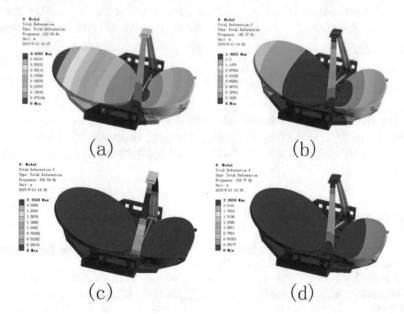

Fig. 10. Modal analysis of the optical system. (a) Shape of the 1 order modal; (b) Shape of the 2 order modal; (c) Shape of the 3 order modal; (d) Shape of the 4 order modal.

4.2.3   The Overload Resistance Analysis of Optical System

Because pod overload for a maximum of 15 g, 15 g of the optical system for three directions of the overload resistance analysis, as shown in Fig. 11.

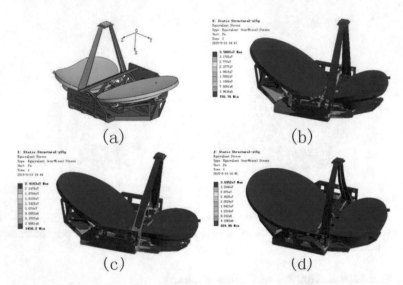

Fig. 11. Overload resistance analysis of the optical system. (a) Definition of the coordinate system; (b) Direction of X; (c) Direction of Y; (d) Direction of Z.

15 g, therefore, three directions under the maximum stress is less than the yield strength, material strength meet the requirement of overload, and has a large safety margin.

5   Conclusion

Using the free-form surface in off-axis reflective imaging system can significantly improve the performance of optical and mechanical structure. We should do more research in popularizing of the free-form surface apply to the space optical imaging systems.

References

1. Li, L., Huang, Y., Wang, Y.: Modern Optical Design, 2nd edn. Beijing Institute of Technology Press, Beijing (2015). (in Chinese)
2. Thompson, K.P.: Description of the third-order optical aberrations of near-circular pupil optical systems without symmetry. JOSA A **22**(7), 1389–1401 (2005)
3. Thompson, K.P.: Multinodal fifth-order optical aberrations of optical systems without rotational symmetry: spherical aberration. JOSA A **26**(5), 1090–1100 (2009)

4. Thompson, K.P.: Multinodal fifth-order optical aberrations of optical systems without rotational symmetry: the comatic aberrations. JOSA A **27**(6), 1490–1504 (2010)
5. Thompson, K.P.: Multinodal fifth-order optical aberrations of optical systems without rotational symmetry: the astigmatic aberrations. JOSA A **28**(5), 821–836 (2011)
6. Fuerschbach, K., Rolland, J.P., Thompson, K.P.: Theory of aberration fields for general optical systems with freeform surfaces. Opt. Express **22**(22), 26585–26606 (2014)
7. Yang, T., Zhu, J., Jin, G.: Nodal aberration properties of coaxial imaging systems using zernike polynomial surfaces. JOSA A **32**(5), 822–836 (2015)
8. Zhu, J., Wu, X., Hou, W., Yang, T., Jin, G.: Application of freeform surfaces in designing off-axis reflective space optical imaging systems. Spacecraft Recov. Remote Sens. (2016) (in Chinese)

Fabrication, Surface Shape Detection and System Error Correction of Precise Large Diameter Convex Aspheric Surface

Yan Du[(✉)], Ji-you Zhang, Peng Wang, Xiao-hui Meng,
and Hui-jun Wang

Beijing Institute of Space Mechanics and Electricity Optical Ultraprecise
Processing Technology Innovation Center for Science and Technology Industry
of National Defense (Advanced Manufacture), Beijing 100094, China
duyan622@163.com

Abstract. Convex aspheric mirrors are core optical elements in space optical systems, and their surface shape errors directly affect the imaging quality of the system. In order to improve the surface accuracy of large diameter convex aspheric mirrors, the high-precision fabrication and testing techniques of large diameter convex aspheric mirrors are studied. With an off-axis three mirror optical system secondary mirror as the optical parameters of convex aspheric mirror, use high precision aspheric surface milling-robot grinding and polishing process-ion beam polishing as machining method. In the grinding stage, the surface shape is measured by three-coordinate measuring instrument, in the polishing stage, the surface shape is measured by meniscus lens with interferometer, the calibration and correction of the meniscus are carried out by using standard spherical mirror. Based on the theoretical analysis of the detection method of aspheric surface, combined with experiments, extracted the error factor. Through the process method, the error parameter of high sensitivity are controlled, and the perfect large aperture convex aspheric test error separation is constructed. The system error is calibrated and the surface distribution of the real aspheric optical component is calibrated. The testing result indicate that the surface figure accuracy are 0.017λRMS. Through this measuring method, full aperture testing of optical aspheric can be realized. All the specifications of the convex aspheric mirror can meet the requirements of the optical design.

Keywords: ISSOIA 2017 · Convex aspheric surface fabrication
Convex aspheric surface shape detection · Surface figure error

1 Introduction

Application of aspheric surface in optical system can simplify the structure, improve performance of the system. So, a large number of aspheric surface elements are used in the fields of national defense, aerospace and astronomical observation field, off-axis three mirror optical system has become the most common type of optical system of space camera. With the increasing resolution of the system, the aperture of optical system is becoming larger and larger. It is inevitable to use large aperture convex

© Springer Nature Switzerland AG 2018
H. P. Urbach and Q. Yu (Eds.): ISSOIA 2017, SPPHY 209, pp. 134–143, 2018.
https://doi.org/10.1007/978-3-319-96707-3_15

aspheric mirror as the main optical component. Compared with concave aspheric mirror, the fabrication of large diameter convex aspheric surface is very difficult, which is the main factor that restricts the wide application of high precision large aperture convex aspheric optical component. Testing is the key to fabrication, therefore, it is important to study the testing technology of large diameter convex aspheric surface [1].

In recent years, many new space cameras, secondary mirrors have been tested with new hindle shell methods to test convex aspheric surfaces. Hindle shell methods using a concentric low diopter meniscus lens to replace the standard spherical mirror, concave meniscus lens is plated with a layer of semi reverse semipermeable membrane, while allowing light through the realization of the light reflecting function. Because the distance between the aspheric surface and the meniscus lens can be controlled very closely, not only can the meniscus lens size be greatly reduced, but also the convex aspheric surface can be detected with a full aperture. Since the optical path is reflected two times on the secondary mirrors, the accuracy of the surface profile of the secondary mirror is doubled [2–4].

For the large aperture convex aspheric surface processing difficulty, the existing processing and testing equipment as the basis, combined with the engineering application of a Φ450 mm secondary mirror test, optimization of NC machining process existing, used a high precision aspheric surface milling-robot grinding and polishing process-ion beam polishing as machining method. Meniscus lens and standard spherical mirror is designed to delete the system error, test processing in this paper is suitable for large diameter convex aspheric surface, can significantly improve the machining and precision.

2   Testing System Design

The role of concave standard spherical mirror as an alternative, the technical key is to strictly control the radius of curvature is used to conjugate imaging, convex function is to eliminate spherical aberration, the key technique is to strictly control the surface shape accuracy. The concave surface should be deposited on the semi inverse error of semipermeable membrane. The meniscus lens need standard spherical mirror for error correction, which in the system test as error deduction (Fig. 1).

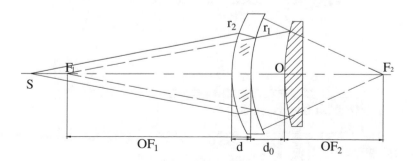

Fig. 1. Meniscus lens testing method [5]

The optical path in the meniscus lens after two times of refraction, a reflection of the system, the contrast will be relatively low, in order to obtain the interference pattern of high contrast, need film reflectivity on the deposition of concave is determined by theoretical calculation, when the concave plating transmittance and reflectance is 66.6%, fringe contrast the best.

2.1   Convex Aspheric Mirror Parameters

The convex aspheric mirror designed in the development of a space optical camera project has an effective aperture of $\Phi450$ mm and a hyperboloid of the surface. R = −808.74 mm, K = −2.174.

$$z = cy^2 / \left(1 + \sqrt{1 - (1 + K)c^2 y^2}\right) \tag{1}$$

$$c = 1/R \tag{2}$$

2.2   Optical Path Design of Meniscus Lens

The light source is placed near the focus of the convex aspheric surface, the initial structure is calculated, and the data is inserted into the optical simulation software. The optical path parameters are obtained by using the software to optimize the initial structure.

The residual aberration of the meniscus lens detection system is PV = 0.0003λ, RMS = 0.0001λ (λ = 632.8 nm): in which the center thickness of meniscus lens is 80 mm (Figs. 2 and 3).

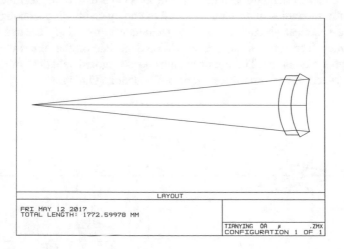

Fig. 2. Testing layout of meniscus lens

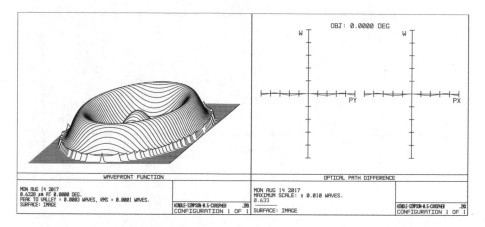

Fig. 3. Residual wavefront errors and opd fan

3   System Error Correction

Since the final surface quality of the aspheric mirror is subject to the manufacturing accuracy of the meniscus lens, a detailed tolerance analysis of the compensator must be carried out. The main error sources that affect the detection system include system design residual wavefront errors, system distortion errors, concave wavefront error of the meniscus lens, convex wavefront error of the meniscus lens, inhomogeneous error of meniscus lens material, interferometer lens error.

3.1   Convex Wavefront Error of the Meniscus Lens and Inhomogeneous Error of Meniscus Lens Material Correction

Place the point source in the secondary mirror geometry focus, light rays pass through the meniscus lens and concave mirror is formed without aberration detection, the same principle, place a standard spherical mirror lens in the rear, in the vicinity of the geometrical focus can always find a place to meet the self collimation detection. By means of optical design software, multi-structure design method is adopted to optimize the design of auto-collimating light path while taking into account the meniscus lens to detect convex aspheric optical path. The error of convex surface error and material nonuniformity error of meniscus lens is removed by self collimation detection light path (Figs. 4 and 5).

The standard lens is a concave mirror, a point light source to meniscus lens distance is R = 1850.724 mm, two optical detection point source to the meniscus lens distance is completely consistent, so as to eliminate error of meniscus lens detection system, design standard spherical mirror detection system for residual aberration. The residual aberration of the meniscus lens detection system is PV = 0.0183λ, RMS = 0.0042λ (λ = 632.8 nm)

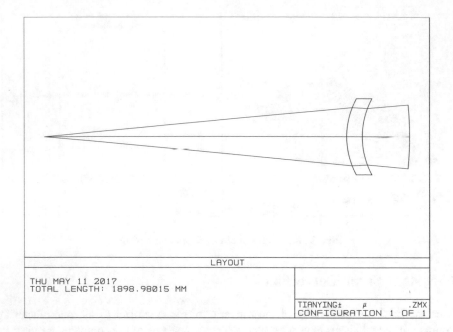

Fig. 4. System error correction layout of meniscus lens

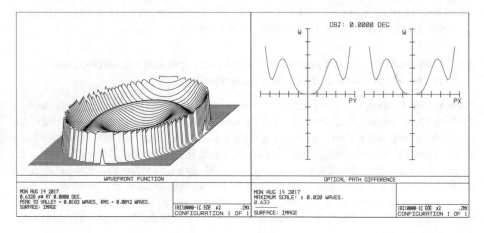

Fig. 5. Residual wavefront errors and opd fan

3.2 System Distortion Correction

As a high precision surface testing device, the interferometer can realize the mapping of CCD pixel coordinates to the actual specular coordinates at the same time, so that the processing equipment can work in the correct space position. Type test mainly though PZT image motion or wavelength shifting like, pixel coordinates mapping to the actual coordinates have to be mirror imaging is realized by interferometer.

The current use of commercial phase-shifting interferometer field is relatively small, and the imaging system can be directly used for small aberrations, zero detection of spherical and plane, for the detection of aspheric mirror requires compensation mirror as the compensator, the interference of spherical wave conversion process instrument and tested, non spherical surface non spherical wave. The design for a long time because all degrees of freedom for phase compensation of the compensator, so there is big aberration compensator itself, although the phase shifting interference contrast instrument itself is very small, but when the compensator in the test optical path of the interference, there will be a lot of aberration, if in the working state of the eccentric compensator system result is used to obtain the mapping relation between the pixel coordinates and the actual coordinates of the mirror

In order to complete the correction of the distortion of the system, as the grid market interval in aspheric surface testing is required, the manual processing method to find the interference point coordinates pixel labeled graph, obtained two mapping between CCD pixel coordinates and real surface coordinates. Through the optical design software, the system is compensated for real ray tracing, the distortion value is calculated. And the robot software is used to guide the processing and eliminate the system distortion [6].

The nonlinear model and nonlinear error of the meniscus lens are established, and the normalized aperture value of interferometer CCD is defined as h_p, the normalized aperture value of the actual mirror is defined as h_m, there is no nonlinear error at the edge of the mirror, so h_p is equal to h_m, In other positions of the mirror, the nonlinear errors are different, and the derivative is zero when the derivative of $h_m - h_p$ is obtained. By calculation, the nonlinear error is most sensitive to the 0.577 ring zone of the reflector, and there are two ways to calculate the 0.577 ring zone distortion. The first method is through the optical design software, the system is compensated for real ray tracing, the distortion value is calculated. And the robot software is used to guide the processing and eliminate the system distortion. The second method is the method of time measurement. The interference fringes are judged by marking the secondary mirrors, and the values obtained by actual measurement are compared with the actual ray tracing results (Fig. 6).

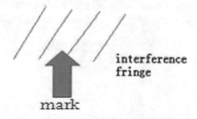

Fig. 6. Actual measurement distortion

3.3   Interferometer Lens Error Removal

In order to carry out high-precision measurement of the secondary mirror, it is necessary to calibrate the interferometer standard lens. The interferometer lens error file is manufactured by calibrating the interferometer standard lens error removal. The error of the test result is subtracted to obtain the ture sub mirror shape error.

4   Experimental Verification

4.1   Machining Process

Large aperture convex aspheric machining process consists of high precision aspheric surface milling-robot grinding and polishing-ion beam polishing in three stages, the aspheric milling back shape precision can reach 1umRMS, robot grinding and polishing surface precision can reach λ/30 RMS, ion beam polishing can reach λ/70RMS, Robot grinding and polishing Equipment based on CCOS polishing principle. An intelligent numerical control system with six axes and multiple degrees of freedom is adopted. The end of the flexible lapping tool is perpendicular to the normal direction of the mirror (Fig. 7).

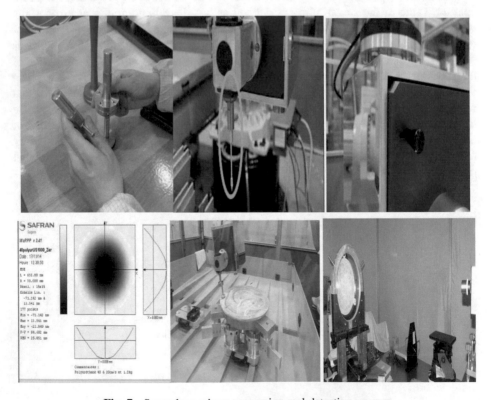

Fig. 7. Secondary mirror processing and detection process

4.2 Date Processing

After the machining, the concave surface of the meniscus lens is RMS = 0.010λ, the shape accuracy of standard spherical mirror surface is RMS = 0.008λ. The shape accuracy of secondary mirror is RMS = 0.023, the test results are shown in Fig. 8 below. Through the standard spherical mirror, the meniscus, convex surface and material inhomogeneity of the meniscus are removed, and error removal files are obtained. Error removal files are shown in Fig. 9 below. Date processing is carried out by combining Figs. 9 and 10 to obtain the true surface shape test result of the secondary mirror. The shape accuracy of secondary mirror is RMS = 0.016λ.

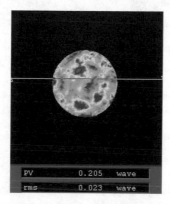

Fig. 8. Original interferogram

4.3 Error Analysis

The wavefront errors of the secondary mirror detection system include meniscus lens system design residual error w_1, standard spherical mirror system design residual error w_2, system distortion errors w_3, optical path adjustment error w_4, concave wavefront error of the meniscus lens w_5, convex wavefront error of the meniscus lens and inhomogeneous error of meniscus lens material w_6, interferometer lens error w_7, wavefront error of standard spherical mirror w_8. By correction, system distortion errors w_3, optical path adjustment error w_4, convex wavefront error of the meniscus lens and inhomogeneous error of meniscus lens material w_6, interferometer lens error w_7 not affect the wavefront of the detection system. According to optical design results meniscus lens system design residual error $w_1 = 0.0001\lambda$, standard spherical mirror system design residual error $w_2 = 0.0042\lambda$, The light is refracted two times at the concave of the meniscus, so its wavefront error is $2 \times (n - 1)$ times so the concave wavefront error of the meniscus lens $w_5 = 2 \times (n - 1) \times 0.01\lambda = 0.01\lambda$. The light is reflected at the standard spherical mirror once, so its wavefront error $w_8 = 2 \times 1.5 \times 0.0008\lambda = 0.0024\lambda$.

According to the principle of RRS error synthesis [7], the overall wavefront error of the detection system can be expressed as

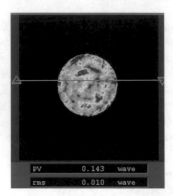

Fig. 9. Error removal files

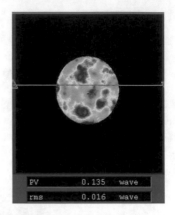

Fig. 10. Final interferogram

$$w = \sqrt{w_1^2 + w_2^2 + w_3^2 + w_4^2 + w_5^2 + w_6^2 + w_7^2 + w_8^2}$$
$$= \sqrt{0.0001\lambda^2 + 0.0042\lambda^2 + 0 + 0 + 0.01\lambda^2 + 0 + 0 + 0.0024\lambda^2}$$
$$\approx 0.0111\lambda$$

According to 3σ detection principle, the result can meet the design requirements of the system [7]. Finally, the machined surface accuracy of the Φ450 mm convex aspheric mirror is PV = 0.165, RMS = 0.016λ, λ = 632.8 nm.

5 Conclusion

This paper Φ450 mm large aperture convex aspheric mirror were detected by processing, high precision production line of advanced, combined with the design and calibration of optical mirror system error distortion calibration method to remove the system error, in the process to precisely control the geometric parameters of the convex aspheric machining surface to PV = 0.165λ, RMS = 0.016λ, λ = 632.8 nm. the result meet the design requirements of the system, the process and the method can detect the surface shape of the high precision test of full aperture convex aspheric surface, can be extended to large convex aspheric surface full aperture testing process, has a broad application prospect.

References

1. Wang, X.K.: Measurement of convex aspheric mirror by non-null testing. J. Appl. Opt. **33**(1), 125–126 (2012)
2. Gao, S.T., Wang, G.W., Zhang, J., et al.: Correction of distortion in asphere testing with computer generated hologram. Opt. Precis. Eng. **8**(2), 371–389 (2013)
3. Yuan, L.J., Chen, T.: Manufacturing technology for high order aspheric surface. J. Appl. Opt. (2011)
4. Feng, Z.: Fabrication and testing of precise off-axis convex aspheric mirror. Opt. Precis. Eng. **12**(2), 2558–2563 (2011)
5. Jie, M.A., Zheng, Z.: Testing convex aspherical surfaces with optimized modified handle arrangement. Infrared Laser Eng. **40**(2), 277–281 (2011)
6. Tsaj, M.J., Huang, J.F., Kao, W.L.: Robotic polishing of precision molds with uniform material removal control. Int. J. Mach. Tools Manuf. **49**(5), 885–895 (2009)
7. Pan, I.H.: The Design, Manufacture and Test of the Aspherical Optical Surface. Suzhou University Press, Suzhou (2004)

Overview of High-Precision Spectral Payload On-Orbit Radiometric Calibration

Shaofan Tang[✉], Long Chen, Huan Li, and Zhi Jun Lu

Beijing Institute of Space Mechanics and Electricity, Key Laboratory for Advanced Optical Remote Sensing Technology of Beijing, No. 104, Youyi Road, Haidian District, Beijing 100094, China
sftang508@sina.com

Abstract. High-precision spectral payload during the track performance changes occur, and there for need to be calibrated by on-orbit radiation calibration technology. This paper describes several types of on-orbit radiation calibration techniques. On the basis of summarizing the typical application of the international on-orbit calibration technology. Aiming at the large-caliber spectral load, this paper proposes an on-orbit scale calibration scheme combined with "full aperture part of the optical path calibration" and "all optical path part of the aperture calibration".

Keywords: High-precision · Large-caliber · Spectral payload
On-orbit radiation calibration · Diffuse reflector

1 Introduction

Spectral imaging is an optical technique that combines spatial optical imaging with spectroscopy and radiometry to produce images. The application of high-precision spectral payloads is based on quantitative data and therefore requires accurate calibration. During the launch and rail operation, the optical, structural and electronic components of the Spaceborne Imaging Spectrometer will change, this leads to the changes between the digitized input established by the laboratory radiometric calibration and the radiance of the ground surface, In order to obtain accurate spectral image data, these changes must be corrected, so the spectral payload not only requires laboratory calibration also need to be on the on-orbit radiation calibration.

2 On-Orbit Radiometric Calibration

On-orbit radiation calibration technology mainly includes site calibration, cross calibration, moon calibration and on-orbit calibration, these methods can be absolute radiation on the remote sensor calibration.

Site calibration, cross calibration, moon calibration and other methods, because of its venues, weather and other factors, is generally difficult to achieve fast, normal on-orbit calibration. Therefore, in order to achieve on-orbit fast normalization of the calibration, spectral load must be equipped with on-orbit calibration device. In the

© Springer Nature Switzerland AG 2018
H. P. Urbach and Q. Yu (Eds.): ISSOIA 2017, SPPHY 209, pp. 144–149, 2018.
https://doi.org/10.1007/978-3-319-96707-3_16

radiometric calibration model of on-orbit calibration, the on-orbit solar radiator calibration method is a widely recognized high precision absolute radiation calibration
method. Using the accuracy of the solar radiation model and high stability of the diffuse
reflector reflection performance, we can pass the least transmission link to get a higher
radiation calibration accuracy, MODIS, MERIS are using diffuse reflector to Perform
on-orbit calibration.

3 Standard Application of On-Orbit Calibration Technology

3.1 MODIS

MODIS on-orbit calibration equipment includes a solar diffuse reflector, solar diffuser
stable monitor, a blackbody, a spectral radiation calibration equipment. In addition,
when the field of view of the remote sensor passes through the spatial observation area,
the space is measured to obtain a zero signal reference. During the orbit, the long
diffuse sunlight is usually aging. Thus in the design phase, MODIS tracks the attenuation of the two-way reflection factor of the solar diffuser by installing a solar diffuser
stable monitor. The solar diffuse attenuation coefficient of all solar reflection bands is
obtained by comparing the attenuation coefficients of the solar diffuse reflectors
measured at nine different wavelengths by using the solar diffuser stable monitor, using
the solar diffuse attenuation coefficient we can calculate the scaling factor of the solar
reflection band. Since the Sun diffuse reflector monitor uses the same set of optical
systems, the detection element and the electronic system observe the sunlight and the
diffuse reflectance of the solar diffuse reflector, Even if the detection factor of the
detection element, the optical transmittance, or the stability of the electronic system of
the solar diffuser stable monitor changes slowly, it will not affect the quality of the solar
diffuse reflector attenuation monitoring results (Fig. 1).

3.2 EnviSat-1 MERIS

The medium resolution imaging spectrometer (MERIS) on the European environmental
satellite EnviSat-1 is mainly used for ocean monitoring. When the satellite flew over
the Antarctic, and the sun is perpendicular to the direction of the sky, the solar diffuse
reflector reflects the sunlight into the field of view, thus achieving the full aperture, and
the same pattern of observation (Fig. 2). The calibration mechanism can choose five
positions to realize the dark current calibration, the ground observation, the radiation
calibration, the diffuse reflector attenuation characteristic monitoring and the spectral
calibration. The diffuse reflectance of the diffuse reflectance is analyzed by using
another diffuse reflector plate that is the same as the radiation diffuser. By periodically
radiating the calibration, the signals of the two diffuse reflectors and the variation of the
bidirectional reflection function of the main diffuse reflector are compared. With this
information, the degree of attenuation of the diffuse reflector can be evaluated.

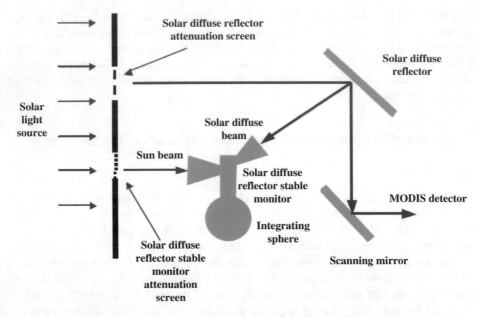

Fig. 1. Solar diffuser stable monitor for MODIS solar calibration

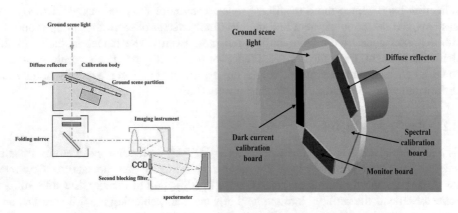

Fig. 2. Sketch map of MERIS on-orbit radiation calibration

3.3 Landsat-8/LDCM

Landsat-8 is the latest generation of landed satellites launched in 2013, this satellite is also known as the land Satellite Data Continuous Mission (LDCM). Its main load OLI's calibration system operates under two calibration modes, the first calibration method (Figs. 3 and 4): By using a rotatable calibration mechanism, radiometric calibration, attenuation coefficient monitoring of diffuse reflector and remote sensor imaging can be realized. The second calibration method stops the external radiation of the remote sensor from entering the optical system by shutting off the shutter, then we use two calibration lamp to calibrate the remote sensor.

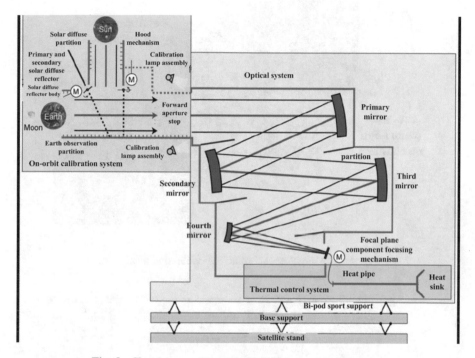

Fig. 3. Sketch map of landsat-8/LDCM on-orbit calibration

4 Design of Large Caliber Spectral Load On-Orbit Calibration Technology

The current on-orbit calibration device (visible near infrared spectrum) commonly used as the sun as an external reference standard source, most of the program using the sun + diffuse transmission plate + solar diffuser stable monitor + diffuse transmission plate group calibration technology, With the use of large-diameter spectral payload more and more widely, the size and uniformity of the solar diffuse transmission plate has been difficult to meet the requirements, In view of the above problems, take GEO-OCULUS payload as reference this paper presents a "full aperture part of the optical path calibration" + "all optical path part of the aperture calibration" combination of on-orbit calibration program, the sun as a standard light source to achieve high precision on-line calibration.

As shown in Fig. 5, large caliber telescope load system calibration subsystem consists of three parts: all optical path part of the aperture calibration equipment, full aperture part of the optical path calibration equipment, Calibration of incident light entrance.

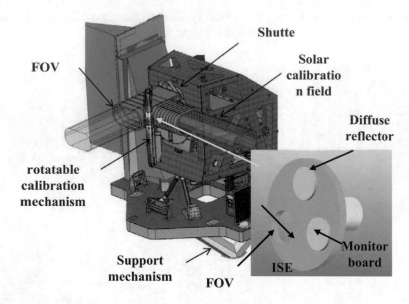

Fig. 4. OLI structure and rotatable calibration mechanism

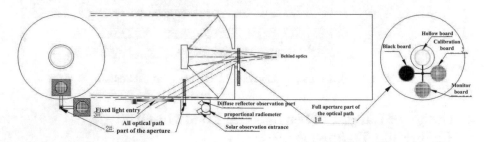

Fig. 5. Sketch map of large-caliber spectral payload on-orbit calibration

Full aperture part of the optical path calibration equipment using double-plate (calibration board + monitor board) calibration and monitoring methods, and set a blackboard for dark current measurement. All optical path part of the aperture calibration installed in the telescope second mirror, easier to introduce sunlight to observe the sun, using a radiometer as the monitoring method.

5 Conclusion

This paper first introduces several kinds of on-orbit radiation calibration techniques. The on-orbit calibration device is becoming an indispensable part of the high-precision spectral payload due to on track fast, normalized calibration. And then briefly sort out some of the international application of on-orbit calibration technology. The current on-orbit calibration device commonly used as the sun as an external reference standard source, most of the program using the sun + diffuse transmission plate + solar diffuser stable monitor + diffuse transmission plate group calibration technology, Finally, according to the current requirements of large-caliber spectral load, this paper presents a combination of "full aperture part of the optical path calibration" + "all optical path part of the aperture calibration" in-orbit calibration scheme.

References

1. Mendenhall, J.A., Ryan-Howard, D.P.: Earth Observing1 Advanced Land Imager: Spectral Response Calibration. Howard, Washington (2000)
2. Markham, B.L., Helder, D.L.: Forty-year calibrated record of earth-reflected radiance from Landsat: a review. Remote Sens. Environ. **122**(3), 30–40 (2012)
3. Slater, P.N., Biggar, S.F., Holm, R.G., et al.: Reflectance- and radiance-based methods for the in-flight absolute calibration of multispectral sensors. Remote Sens. Environ. **22**(1), 11–37 (1987)
4. Chander, G., Markham, B.: Revised Landsat-5 TM radiometric calibration procedures and postcalibration dynamic ranges. IEEE Trans. Geosci. Remote Sens. **41**(11), 2674–2677 (2003)
5. Roithmayr, C.M., Lukashin, C., Speth, P.W., et al.: Opportunities to intercalibrate radiometric sensors from international space station. J. Atmos. Oceanic Technol. **31**(4), 890–902 (2014)
6. Vermote, E., Santer, R., Deschamps, P.Y., et al.: In-flight calibration of large field of view sensors at short wavelengths using Rayleigh scattering. Int. J. Remote Sens. **13**(18), 3409–3429 (1992)
7. Malaguti, G., Bazzano, A., Bird, A.J., et al.: In-flight calibrations of IBIS/PICsIT. Astron. Astrophys. **411**(1), L173 (2003)
8. Xiong, X., Chiang, K., Esposito, J., et al.: MODIS on-orbit calibration and characterization. Metrologia **40**(1), S89 (2003)
9. Xiong, X., Wenny, B., Sun, J., et al.: Status of MODIS on-orbit calibration and characterization. Proc. SPIE **8889**(6), 4128–4131 (2004)
10. Lee, H.J., Liu, Y., Coull, B.A., et al.: A novel calibration approach of MODIS AOD data to predict PM2.5 concentrations. Atmos. Chem. Phys. **11**(11), 9769–9795 (2011)
11. Delwart, S., Preusker, R., Bourg, L., et al.: MERIS in-flight spectral calibration. Int. J. Remote Sens. **28**(3–4), 479–496 (2007)
12. Bourg, L., Delwart, S., Huot, J.P., et al.: Calibration and early results of MERIS on ENVISAT. In: 2002 IEEE International Geoscience and Remote Sensing Symposium, 2002. IGARSS 2002, vol. 1, pp. 599–601. IEEE (2003)
13. Mishra, N., Haque, M., Leigh, L., et al.: Radiometric cross calibration of Landsat 8 Operational Land Imager (OLI) and Landsat 7 Enhanced Thematic Mapper Plus (ETM+). Remote Sens. **6**(12), 12619–12638 (2014)
14. Markham, B.L., Dabney, P.W., Murphy-Morris, J.E., et al.: The landsat data continuity mission operational land imager (OLI) radiometric calibration. In: Geoscience and Remote Sensing Symposium, pp. 2283–2286. IEEE (2010)
15. Martínez, D., Silva, G., Solís, I., et al.: The ground-based absolute radiometric calibration of Landsat 8 OLI. Remote Sens. **7**(1), 600–626 (2015)
16. Markham, B., Morfitt, R., Kvaran, G. et al.: OLI Radiometric Calibration (2011)

A Region Dense Matching Algorithm for Remote Sensing Images of Satellite Based on SIFT

Ning Yang$^{(\boxtimes)}$, Fei Shao, Jing-shi Shen, and Yun Jia

Shandong Institute of Space Electronic Technology,
Yantai 264670, People's Republic of China
yn3714@163.com

Abstract. In the acquisition of 3D terrain information based on images of satellite, image matching is one of the crucial issues. Feature based matching can provide robust results; however, the results are always sparse, and cannot satisfy the need of application. To solve this problem, a region dense matching algorithm for remote sensing images of satellite based on SIFT (Scale Invariant Feature Transform) is presented. In the algorithm, robust matching results of SIFT are taken as the basis, and then matching growing is conducted in different regions of satellite image with affine transformation and least square method. Experiment results show that the algorithm can achieve dense matching for remote sensing images of satellite, especially in few textures or no textures regions.

Keywords: Image matching · Region dense matching · SIFT
Affine transformation · Least Square Method (LSM)

1 Introduction

In recent years, extracting 3D information of terrain from satellite images is a hotspot [1–3]. In order to automatically get precise and robust results, image matching is one of the crucial issues [4, 5]. According to the matching primitive, image matching can be divided into two categories [6]: area based matching, and feature based matching. Area based matching is simple, and has high precision; however, it is sensitive to the radiation distortion, and difficult to choose the size of matching window. Feature based matching has certain resistance to image rotation and scale transform, and can get robust matching results. As a result, feature based matching is widely used, and numerous methods have been proposed, in which SIFT method proposed by D. G. Lowe in 2004 is one of the representative methods [7].

Although feature based matching can provide robust results, for the amount of features extracted from image is limited, the matching results are always sparse, which cannot satisfy the requirements of some applications, such as the extraction of terrain 3D information from satellite images. To solve this problem, a region dense matching algorithm for remote sensing images of satellite based on SIFT is presented. In the algorithm, robust matching results of SIFT method are taken as the basis, and then matching growing is conducted in different regions of satellite image with affine

© Springer Nature Switzerland AG 2018
H. P. Urbach and Q. Yu (Eds.): ISSOIA 2017, SPPHY 209, pp. 150–159, 2018.
https://doi.org/10.1007/978-3-319-96707-3_17

transformation and least square method. Finally, dense matching results are generated so as to achieve the high precision extraction of terrain 3D information from satellite images.

2 Feature Matching Based on SIFT

SIFT method is first taken to conduct initial matching, which mainly includes two parts: detection of key point and construction of local descriptor [8, 9]. Firstly, scale space is generated and feature points are detected and extracted, and then key points are accurately located. Next, direction parameters are specified for each key point, and the descriptor of key point is generated so as to complete feature matching between images.

2.1 Detection of Key Point

The purpose of scale space theory is to simulate multi-scale characteristics of image data. Gaussian convolution kernel is the only linear nucleus to realize the scale change. The scale space of a 2D image is defined as [10]

$$L(x, y, \sigma) = G(x, y, \sigma) * I(x, y), \tag{1}$$

where $I(x, y)$ indicates the 2D image, and $G(x, y, \sigma)$ is a scale-variable Gaussian function expressed as

$$G(x, y, \sigma) = \frac{1}{2\pi\sigma^2} e^{-(x^2 + y^2)/2\sigma^2}, \tag{2}$$

in which (x, y) expresses the location, and σ is the scale parameter. The value of σ determines the smoothness of the image. Here, feature points are detected in difference of Gaussians scale space, which can be constructed as

$$\begin{aligned} D(x, y, \sigma) &= (G(x, y, k\sigma) - G(x, y, \sigma)) * I(x, y) \\ &= L(x, y, k\sigma) - L(x, y, \sigma) \end{aligned} \tag{3}$$

SIFT feature point is the point with local extreme value. In order to find it in scale space, each sample point is compared with 18 points in the adjacent scales and 8 points in the same scale. If the value of sample point is greater or less than that of above 26 points, it is taken as the candidate key point, as shown in Fig. 1.

The candidate key points are identified in above section. In order to improve noise immunity and enhance the stability of matching, points with low contrast and unstable edge response points should be removed.

The Taylor expansion of the scale space function, $D(x, y, \sigma)$, at the sample point is as follows

$$D(\boldsymbol{x}) = D + \left(\frac{\partial D}{\partial \boldsymbol{x}}\right)^{\mathrm{T}} \boldsymbol{x} + \frac{1}{2} \boldsymbol{x}^{\mathrm{T}} \frac{\partial^2 D}{\partial \boldsymbol{x}^2} \boldsymbol{x}, \tag{4}$$

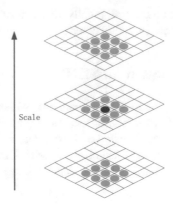

Fig. 1. Detection of point with local extreme value in scale space

where D and its derivatives are evaluated at the sample point and $x = (x, y, \sigma)^{\mathrm{T}}$ is the offset from this point. The location of the extremum, $\hat{x}$, is determined by taking the derivative of this function with respect to x and setting it to zero, giving

$$\hat{x} = -\left(\frac{\partial^2 D}{\partial x^2}\right)^{-1} \frac{\partial D}{\partial x}. \tag{5}$$

The function value at the extremum, $D(\hat{x})$, is useful for rejecting the candidate key points with low contrast. This can be obtained by substituting (5) into (4), giving

$$D(\hat{x}) = D + \frac{1}{2}\left(\frac{\partial D}{\partial x}\right)^{\mathrm{T}} \hat{x}. \tag{6}$$

All extrema with a value of $|D(\hat{x})|$ less than 0.03 are discarded.

For stability, it is not sufficient to reject key points with low contrast and points along an edge should also be eliminated with the property of Hessian matrix H. We only need to check

$$\frac{Tr(H)^2}{Det(H)} < \frac{(\gamma + 1)^2}{\gamma}. \tag{7}$$

Normally, we use a value of $\gamma = 10$, eliminating key points that have a ratio between the principal curvatures greater than 10.

2.2   Construction of Local Descriptor

By assigning a consistent orientation to each key point based on local image properties, the key point descriptor can be represented relative to this orientation and therefore achieve invariance to image rotation. At the beginning, for each image sample, $L(x, y)$, at this scale, the gradient magnitude, $m(x, y)$ and orientation, $\theta(x, y)$, is precomputed by using pixel differences:

$$
\begin{cases}
m(x, y) = \sqrt{(L(x+1, y) - L(x-1, y))^2 + (L(x, y+1) - L(x, y-1))^2} \\
\theta(x, y) = \tan^{-1}((L(x, y+1) - L(x, y-1))/(L(x+1, y) - L(x-1, y)))
\end{cases}
\tag{8}
$$

Then an orientation histogram is formed from the gradient orientations of sample points within a region around the key point. The orientation histogram has 36 bins covering the 360° range of orientations. Each sample added to the histogram is weighted by its gradient magnitude and by a Gaussian-weighted circular window with a σ, which is 1.5 times of the scale of the key point.

Peaks in the orientation histogram correspond to dominant directions of local gradients. The highest peak in the histogram is detected as the orientation of the key point, and then any other local peak that is within 80% of the highest peak is used to also create a key point with that orientation, which contribute significantly to the stability of matching, as shown in Fig. 2.

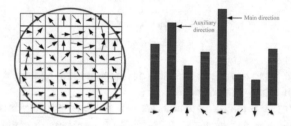

Fig. 2. Estimation of the orientation of the key point

In order to achieve orientation invariance, the coordinates of the descriptor and the gradient orientations are firstly rotated relative to the key point orientation. The descriptor for each key point is then created based on a patch of pixels in its neighborhood. Magnitude and orientation of the gradient are computed for each pixel in a patch of 16 × 16 around the key point. A Gaussian weighting function is used to assign a weight to each magnitude. Then magnitudes with their weights are accumulated into orientation histograms that have 8 directions summarizing the contents over 4 × 4 subregions, as shown in Fig. 3. At last we obtain a 128-dimension vector (SIFT descriptor) for each key point. This 128-element vector is then normalized to unit length so as to make it invariant over changes in illumination.

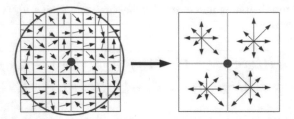

Fig. 3. Construction of the local descriptor

3   Region Dense Matching Based on Affine Transformation

Once matching results of SIFT are acquired, the region with few matching points is selected. Then dense matching is conducted in the region based on affine transformation. The specific steps are as follows:

(1) In order to calculate initial parameters of affine transformation, at least 3 pairs of matching points are selected in the region, whose distribution should be as uniform as possible. The general form of affine transformation for 2D image is as [11]

$$\begin{cases} x' = a_0 + a_1 x + a_2 y \\ y' = b_0 + b_1 x + b_2 y \end{cases} \tag{9}$$

where (x', y') and (x, y) are the coordinates of image points, and $a_i, b_i, i = 0, \cdots, 2$ are the parameters of affine transformation.

According to (9), a matrix equation $AX = b$ is obtained, in which A is the coefficient matrix, and $X = [a_0, a_1, a_2, b_0, b_1, b_2]^{\mathrm{T}}$. With least square method, above matrix equation can be solved as

$$X = (A^{\mathrm{T}}A)^{-1}A^{\mathrm{T}}b. \tag{10}$$

(2) In the left image, select image points that need to be matched, which can be given manually, or selected with grid method. Then coarse matching is conducted in the right image with affine transformation.

(3) Calculate the correlation coefficient ρ of matching points. Take the matching point in the left image as the center, and get an image window with certain size. For each image point in the window, calculate its relative image point in the right image with affine transformation, whose gray value can be obtained by bilinear interpolation. Then the correlation coefficient ρ is calculated as [12]

$$\rho = \frac{\sum\limits_{r=1}^{R}\sum\limits_{c=1}^{C}(g_1(r,c) - \mu_1)(g_2(r,c) - \mu_2)}{\sqrt{\sum\limits_{r=1}^{R}\sum\limits_{c=1}^{C}(g_1(r,c) - \mu_1)^2 \sum\limits_{r=1}^{R}\sum\limits_{c=1}^{C}(g_2(r,c) - \mu_2)^2}}, -1 \le \rho \le 1, \tag{11}$$

where R, C are the sizes of image window, $g_i(r, c), i = 1, 2$ is the gray value of image point, and $\mu_i, i = 1, 2$ is the mean value of image points in the image window.

(4) Decide the relationship between the correlation coefficient ρ and the given threshold T. If ρ is larger than T, take the pair of matching points as the correct result; otherwise, turn to step (5).

(5) Calculate the modified values of parameters of affine transformation with least square method [13]. Linearize the error equation of least square matching, and we get

$$v = c_1 da_0 + c_2 da_1 + c_3 da_2 + c_4 db_0 + c_5 db_1 + c_6 db_2 - \Delta g, \tag{12}$$

where $da_0, da_1, da_2, db_0, db_1, db_2$ are the modified values of parameters of affine transformation, Δg is the difference between left image point and right image point, and the coefficients of the error equation are as follows

$$\left. \begin{aligned} c_1 &= \frac{\partial g_2}{\partial x_2} \frac{\partial x_2}{\partial a_0} = (\dot{g}_2)_x = \dot{g}_x \\ c_2 &= \frac{\partial g_2}{\partial x_2} \frac{\partial x_2}{\partial a_1} = x \dot{g}_x \\ c_3 &= \frac{\partial g_2}{\partial x_2} \frac{\partial x_2}{\partial a_2} = y \dot{g}_x \\ c_4 &= \frac{\partial g_2}{\partial y_2} \frac{\partial y_2}{\partial b_0} = \dot{g}_y \\ c_5 &= \frac{\partial g_2}{\partial y_2} \frac{\partial y_2}{\partial b_1} = x \dot{g}_y \\ c_6 &= \frac{\partial g_2}{\partial y_2} \frac{\partial y_2}{\partial b_2} = y \dot{g}_y \end{aligned} \right\}, \tag{13}$$

where g_2 is the gray value of image point in the right image, $\dot{g}_x$ is the partial derivative of image point in x direction, and $\dot{g}_y$ is the partial derivative of image point in y direction.

In the digital image,

$$\begin{cases} \dot{g}_x = \dot{g}_I(I, J) = \frac{1}{2}[g_2(I+1, J) - g_2(I-1, J)] \\ \dot{g}_y = \dot{g}_J(I, J) = \frac{1}{2}[g_2(I, J+1) - g_2(I, J-1)] \end{cases}. \tag{14}$$

According to (12) and (13), for each image point in the window, construct error equation as

$$V = CX - L, \tag{15}$$

where $X = [da_0, da_1, da_2, db_0, db_1, db_2]^{\mathrm{T}}$.

From (15), we get

$$(C^{\mathrm{T}}C)X = (C^{\mathrm{T}}L), \tag{16}$$

in which X can be solved with least square method.

(6) Modify the parameters of affine transformation with following equation,

$$\begin{bmatrix} 1 & 0 & 0 \\ a_0^i & a_1^i & a_2^i \\ b_0^i & b_1^i & b_2^i \end{bmatrix} = \begin{bmatrix} 1 & 0 & 0 \\ da_0^i & 1+da_1^i & da_2^i \\ db_0^i & db_1^i & 1+db_2^i \end{bmatrix} \begin{bmatrix} 1 & 0 & 0 \\ a_0^{i-1} & a_1^{i-1} & a_2^{i-1} \\ b_0^{i-1} & b_1^{i-1} & b_2^{i-1} \end{bmatrix}, \tag{17}$$

where $a_0^{i-1}, a_1^{i-1}, a_2^{i-1}, b_0^{i-1}, b_1^{i-1}, b_2^{i-1}$ are the last parameters of affine transformation, and $a_0^i, a_1^i, a_2^i, b_0^i, b_1^i, b_2^i$ are the modified parameters of affine transformation.

(7) For the matching point, turn to step (2) to conduct iteration. Only the correlation coefficient ρ is larger than the given threshold T, or reach certain iteration times, stop the iteration.

4 Experiment Results and Analyses

In the experiment, images that to be matched are the forward image and afterward image of IRS-P5 satellite, whose spatial resolution is 2.5 m, as shown in Fig. 4. Matching results of SIFT are shown in Fig. 5.

Fig. 4. Images that to be matched

In the forward image, the region $x \in [10, 435], y \in [275, 515]$ is selected to conduct dense matching, in which only 7 pairs of matching points are obtained with SIFT, as shown in Fig. 6. Then select 10 image points randomly in the region to conduct matching with the region dense matching algorithm presented in this paper. The correlation coefficient ρ is defined as 0.6, and the number of iteration for each image point is 7. Finally all the 10 image points in the forward image are well matched in the afterward image, as shown in Fig. 7.

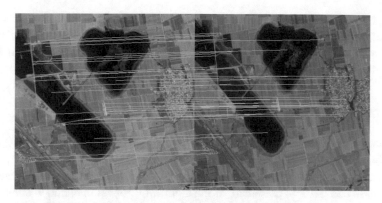

Fig. 5. Matching results of SIFT

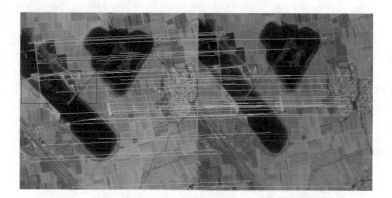

Fig. 6. The region that to be conducted dense matching

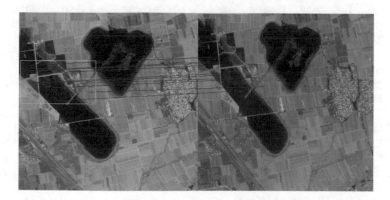

Fig. 7. Dense matching results in the selected region

It can be seen from the experiment results that the algorithm presented in this paper can effectively achieve the dense matching for remote sensing images of satellite, especially for few textures or no textures regions. Furthermore, the algorithm also has certain guidance on solving the matching problem of occlusion area.

5 Conclusion

In the acquisition of 3D terrain information from images of satellite, the results of feature based matching are always sparse, which cannot satisfy the need of application. To solve this problem, a region dense matching algorithm for remote sensing images of satellite based on SIFT is presented. In the algorithm, robust matching results of SIFT are taken as the basis, and then matching growing is conducted in different regions of satellite image with affine transformation and least square method. Experiment results show that the algorithm can achieve dense matching for remote sensing images of satellite, especially for few textures or no textures regions. Furthermore, the algorithm also has certain guidance on solving the matching problem of occlusion area.

It is worth noting that the initial feature based matching directly influence the precision of subsequent dense matching and the convergence of iteration. As a result, how to improve the precision and reliability of initial feature based matching will be the important work in the future.

References

1. Musialski, P., Wonka, P., Aliaga, D.G., Wimmer, M., van Gool, L., Purgathofer, W.: A survey of urban reconstruction. Comput. Graph. Forum **32**, 146–177 (2013)
2. Poli, D., Caravaggi, I.: 3D modeling of large urban areas with stereo VHR satellite imagery: lessons learned. Nat. Hazards **68**, 53–78 (2013)
3. Duan, L., Lafarge, F.: Towards large-scale city reconstruction from satellites. In: 14th European Conference of Computer Vision, Part V, Amsterdam, Netherlands (2016), pp. 89–104
4. Pang, Y., Li, W., Yuan, Y., Pan, J.: Fully affine invariant SURF for image matching. Neurocomputing **85**, 6–10 (2012)
5. Remondino, F., Spera, M.G., Nocerino, E., Menna, F., Nex, F.: State of the art in high density image matching. Photogram. Rec. **29**, 144–166 (2014)
6. Li, Z., Song, L., Xi, J., Guo, Q., Zhu, X., Chen, M.: A stereo matching algorithm based on SIFT feature and homography matrix. Optoelectron. Lett. **11**, 0390–0394 (2015)
7. Lowe, D.G.: Distinctive image features from scale-invariant key points. Int. J. Comput. Vis. **60**, 91–110 (2004)
8. Liu, Y., Liu, S., Wang, Z.: Multi-focus image fusion with dense SIFT. Inf. Fusion **23**, 139–155 (2015)
9. Huo, J., Yang, N., Cao, M., Yang, M.: A reliable algorithm for image matching based on SIFT. J. Harbin Inst. Technol. (New Ser.) **19**, 90–95 (2012)
10. Liang, D., Deng, W., Wang, X., Zhang, Y.: Multivariate image analysis in Gaussian multi-scale space for defect detection. J. Bionic Eng. **6**, 298–305 (2009)

11. Liu, F., Zhou, T., Yang, J.: Geometric affine transformation estimation via correlation filter for visual tracking. Neurocomputing **214**, 109–120 (2016)
12. Mudassar, A.A., Butt, S.: Improved digital image correlation method. Opt. Lasers Eng. **87**, 156–167 (2016)
13. Yang, N., Cheng, Q., Xiao, X., Zhang, L., Jiang, X.: Point cloud optimization method of low-altitude remote sensing image based on vertical patch-based least square matching. J. Appl. Remote Sens. **10**, 035003 (2016)

Filtering Algorithm of False Events in Lightning Detection by FY-4 Lightning Mapping Imager

Huiting Gao[1(✉)], Shulong Bao[1], Wei Liu[1], Hua Liang[1],
Fuxiang Huang[2], and Wen Hui[2]

[1] Beijing Institute of Space Mechanics and Electricity, Key Laboratory
for Advanced Optical Remote Sensing Technology of Beijing,
No. 104, Youyi Road, Haidian District, Beijing 100094, China
gaohuiting_1100@126.com
[2] National Satellite Meteorological Center, China Meteorological
Administration, Beijing 100081, China

Abstract. The lightning detection by Lightning Mapping Imager (LMI) of the
FY-4 geostationary satellite plays a significant role in monitoring strong convection in real time and provides continuous lighting measurements. An
appropriate lightning filtering algorithm is proposed and described in this paper.
The ghost, track and the shot are recognized as the primary non-lightning
artifacts by analyzing the in-board lightning data of the LMI. The ghost is
identified based on the mirror rules of the position and the energy measured in
the laboratory. A line detection method based on the Hough transform is
adopted to eliminate the track. The shot is filtered based on the event clustering
principle. The lightning filter algorithm is applied to process two samples,
sample one is obtained when a strong thunderstorm happened on the south and
the southwest China from March 29th–30th, 2017, the other sample is obtained
on June 21th, 2017 when a short-term serve storm happened in Beijing-Tianjin-
Hebei region. The lightning data obtained by the LMI and synchronous ground
based strokes was processed and compared. The results shows that the spatial
distribution observed from LMI is in general agreement with that of ground-
based monitoring result while the proposed filtering algorithm is applied, providing a well proof of the lightning detected accuracy of LMI.

Keywords: Filtering · Non-lightning artifacts · Hough transform
Clustering

1 Introduction

Lightning is a global phenomenon that has close connection with severe convection
weather such as storms, tornado activity and so on, while it is detected from satellite,
the distribution, variation and position of lightning in large areas can be captured at the
same time, what will aid in forecasting severe storms, tornado activity and convective
weather impacts on aviation safety and efficiency among a number of potential
applications. OTD and LIS are two lightning detected remote sensors carried by polar

© Springer Nature Switzerland AG 2018
H. P. Urbach and Q. Yu (Eds.): ISSOIA 2017, SPPHY 209, pp. 160–172, 2018.
https://doi.org/10.1007/978-3-319-96707-3_18

satellites launched in the 1990s. Abundant lightning data of OTD and LIS has been used to acquire and investigate the distribution and variability of total lightning over the Earth, and to increase understanding of underlying and interrelated processes in the Earth/atmosphere system [1]. GLM of GOES-R launched in 2016 can map total lightning activity over Americas and adjacent oceanic regions. Four Imaging Satellites of Meteosat Third Generation (MTG) with the Lightning Imager (LI) as payloads will be successively launched from 2021, for detecting, monitoring, tracking and extrapolating, in time, the development of active convective areas and storm life cycles.

LMI has been deployed as an instrument of Chinese FY-4 geostationary satellite. It has continuous monitoring capability across a near fan-shaped coverage with the swath width of 3200 km along the direction of the longitude and 4800 km along the direction of the latitude. LMI operates continuously day and night with near-uniform spatial resolution of 7.8 km over the China and adjacent regions. As shown in Fig. 1, the field-of-view coverage of LMI faces toward northern hemisphere(red curve) from March to September but southern hemisphere(green curve) at other time of each year. LMI employs a narrowband interference filter centered near 777.4 nm, which is near a prominent oxygen emission line triplet in the lightning spectrum. The CCD pixel array images its field of view every 2 ms. A modified frame-to-frame background subtraction is implemented at each pixel to remove the slowly varying background signal from the raw data steam [2]. The mission objectives of LMI are: (1) provide continuous lighting measurements like a television camera for storm warning and nowcasting and (2) accumulate a long-term database to track decadal changes of lightning around China [3].

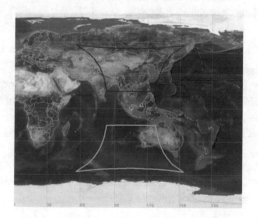

Fig. 1. Field-of-view coverage of LMI

LMI transforms the lightning signal into the continuous image sequence through the optical imaging technology. When a peak exceeds the level of the variable threshold, it is considered to be a lightning event, the lightning data stream contains both real signals and non-lightning artifacts. Hereafter, these non-lightning artifacts are designated as NLEs (Non-Lightning Events), overall rate of the NLES for the lightning data set is about 80%–90% [4]. NLEs can be divided into two kinds based on the cause, internal noise and external noise. Internal noise comprises ghost, shot, S/C jitter and lollypop, external noise comprises cloud radiation, track and sun glint. We have found the shot, track, and ghost noise in the data of LMI.

In addition to on-orbit event detection, the data needs to be filtered to minimize the number of non-lightning events [5, 6]. Many of the artifacts are due to phenomena that occur at the CCD array. In most cases, these artifacts can be removed by event based filtering before the data is geo-located. The remaining artifacts are then filtered based on the local temporal and spatial characteristics of lightning compared with the characteristics of the noise artifact being filtered. This filtering is done after the data has been clustered into groups.

2   Filtering Algorithm of Non-lightning Events

2.1   Definition

The lightning data is composed of the event, which is the basic unit of the lightning. An event denotes a pixel with the DN larger than the threshold value, it may be a discharge of the real lightning or non-lightning signal caused by internal or external reasons. Because a single pixel will almost never correspond to the exact cloud illumination area, a lightning discharge will often illuminate more than one pixel during a single integration time. Two or more adjacent events in the same frame are named as a group. A lightning flash consists of one to multiple optical pulses that occur in the same storm cell within a specified time and distance. A lightning flash should then correspond to several related groups in a limited area [7].

2.2   Ghost

2.2.1   Characteristic of the Ghost

One 1.0-nm-narrow band-pass filter with reflectance of 0.6 is located in front of the optical system of LMI, just like a plane mirror. The stray light of the imaging optical path reflected by the optical component and CCD reaches the filter, then it is reflected by the filter and pass through the optical system, finally assembles in the focal plane to form the ghost image.

Two Laboratory experimental examples of the real signal and the subsequent ghost with different targets are shown in Fig. 2. Note that because of the symmetry of the reflective effect, characteristics of the track are:

(a) the shape of ghost and signal is identical;
(b) the ghost signal is in the "mirror image" location of real signal;
(c) the energy of ghost attenuate proportionally with that of signal.

2.2.2   Description of the Algorithm

The location of the mirror point together with the energy relationship between the signal and the ghost was measured based on the laboratory data, and then will be used to check and the eliminate the ghost. Suppose the position of detected event is (x, y),

Fig. 2. Laboratory examples of signals and ghosts

the position of mirrored pixel (x′, y′) is computed first, the pixel (x′, y′) is ghost while DN values of two pixels is:

$$DN_{(x',y')} < \alpha \cdot DN_{(x,y)} \qquad (1)$$

Where α is the proportional coefficient measured in the laboratory.

Two in-orbit samples of ghost of LMI are shown in Fig. 3. White speckles represent lightning signals, while the red cross denotes ghost. The position and the DN value of events are presented by Table 1. We can find that the energy ratio of ghost to signal is less than five percent, which is accordance with the laboratory measurement result, the relationship of "mirror image" is invariable.

Fig. 3. In-orbit examples of signals and ghosts

Table 1. In-orbit data of position and the DN

	Sample 1			Sample 2		
	Row	Column	DN	Row	Column	DN
Ghost	89	209	33	153	215	30
Lightning events	323	88	3428	256	83	3161
	322	88	3197	255	83	2708
	323	89	1580	256	82	2097
	322	89	1406	255	82	1779

(continued)

Table 1. (*continued*)

	Sample 1			Sample 2		
	Row	Column	DN	Row	Column	DN
	321	88		256	84	1119
	323	87	964	255	84	804
	322	87	938	257	83	716
	324	88	798	257	82	604
	321	89	790	256	81	576
	324	89	568	255	81	485
	324	87	451	257	84	481
	322	90	444	254	83	417
	321	87	417	254	82	364
	323	90	423	256	85	352
	324	90	306	254	84	321

2.3   Track

2.3.1   Characteristic of the Track

The track is caused by collisions of high energy particle such as the ultraviolet, solar wind and so on. If the relative movement happens while the high energy particle collides the CCD array plane, the straight line trace generated during a single frame time is considered as the track. The track filter will identify and remove tracks. A track is identified when the source produces a linear group of pixels in a single time frame. In the natural phenomenon, it is very unlikely that lightning would cause a narrow streak over several hundred kilometers in a single 2 ms CCD time frame. Examples of the track captured by LMI are shown in Fig. 4.

Fig. 4. Examples of track

Characteristics of the track are:

(d) the shape is approximately linear;
(e) events are within the same or adjacent frame;
(f) events are continuous for LIS but discontinuous for LMI.

Characteristic of continuity of LMI is different from that of NASA TRMM-LIS because LMI sent only up to the strongest 120 events in each frame. As shown in Fig. 5, the field-of-view of LMI is divided into 8 areas, data of every area is processed by the respective RTEP(real-time event processor).The upper limit to the output number of events in each area is shown in Table 2, which is depended on the difference of the probability of lightning occurrence according to the history data statistics. It has been proved that most of lightning happens in south China, so the output number of area seven is largest.

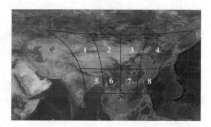

Fig. 5. Distribution of areas and maximum output number of events

Table 2. Upper limit of the output number of events

Maximum output number of events in each area			
Number of area	Count	Number of area	Count
1	6	5	15
2	9	6	15
3	15	7	35
4	10	8	15

2.3.2 Description of the Algorithm

The Hough transform is a technique used in image processing to detect linear features [8]. In general, the transform is applied to image pixels which have already been identified as pixels of interest. Each pixel of interest, or feature pixel, at a given position x, y in an image is mapped into polar coordinate using the normal parameterization of a line:

$$\rho = x\cos\theta + y\sin\theta \tag{2}$$

Where ρ is the distance from the image origin (defined as an image corner) to position x, y and θ is the angle from the horizontal to ρ. Paired with this parameter space is an accumulator array whose size is equal to that of parameter space.

For every feature pixel, pairs of ρ and θ are generated, θ is taken from -90 to $90°$ in a definite degree steps and an ρ value is calculated for each θ value. The representation of the x, y feature pixel in polar coordinate is thus a sinusoidal curve. At every position along the sinusoidal curve in parameter space the accumulator array is incremented by one.

Every feature pixel has a sinusoidal curve in (ρ, θ) parameter space associated with it. Every (ρ, θ) position along the curve represents a series of lines which can go through a particular feature pixel at position x, y. If several feature pixels are part of the same linear feature, their sinusoidal curve representation in polar coordinate will intersect at the (ρ, θ) point which represents the line which goes through all feature pixels. The value of the accumulator array at that particular (ρ, θ) point would be equal to the sum of the sinusoidal curve which intersect at that (ρ, θ) point, or the number of feature pixels on that line. The more sinusoidal curves which intersect, the higher the accumulator array value, and the higher the accumulator array value, the more probable the chance that a linear feature exists in the image. The orientation of the linear feature is given by the value of θ at the accumulator peak location, while the size of the peak is an indication of the size of feature.

In order to eliminate the interference to track detection from the discrete point and the short line by the traditional Hough transformation, some kinds of parameters must be computed first and then used as constraints to avoid the false and missed detection.

2.3.3   Primary Parameters of Hough Transform

The lightning data obtained by LMI during the period from January to March 2017 is used to analyse the primary parameters of Hough transform. The character of the track is as below:

(1) count of the track everyday
     The count of the track is approximately equal every day, from 400 to 500. Events of the track are about 0.7% of the total events.
(2) size of track
     The size of track is used to describe the count of events composing of the track, the histogram of the size is shown in Fig. 6(a) with the maximum of 42 and the minimum of 5, move over 95% of samples is less than 15.

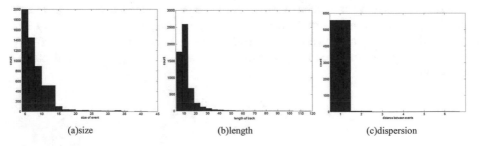

(a)size   (b)length   (c)dispersion

Fig. 6. Histogram of the track

(3) length of track
    The length of track is calculated by the distance between the start and end point, the histogram of the length is shown in Fig. 6(b) with the maximum of 117.8 and the minimum of 5, move over 95.8% of samples is less than 30 (Fig. 6).
(4) The dispersion of events
    The dispersion of events is represented by the distance between the adjacent events of one track, the maximum is 6.7, but more than 98.8% of samples is equal to 1. The dispersion of events is shown in Fig. 6(c).
    Size of track is used as a constraint on the minimal of the accumulator array value. Length of track and dispersion is used for short lines segment elimination and combination.

2.3.4 Filtering Process of Track

The filtering process of track is shown in Fig. 7. First, the threshold vector is constructed by analyzing the amount of in-orbit track samples of LMI, we use the track samples obtained during January to March 2017 at present. Second, by the coordinate transformation from image coordinate to the polar coordinate, every point in the image space maps on a curve. Third, a accumulator is initialized and assigned based on intersections. Next, the maximum value of the accumulator is selected based on the size of track, finally, the length and dispersion are used as constraints to exclude and combine lines.

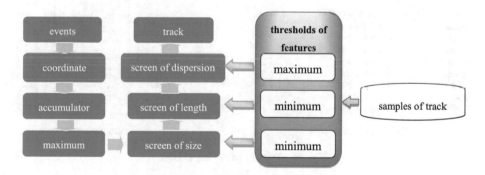

Fig. 7. Workflow of track filter

2.4 Shot

2.4.1 Characteristic of the Shot

Shot noise is the random noise, one kind of shot is caused by the charge fluctuation of the instrument, which is response to a single weak event, another kind of shot is caused by high energy particle collisions, there are only several events while the incident angle of the collision is small enough, finally the pollution of image is may be regard as the shot. It is temporal but without typical features of real lightning such as the time and spatial coherency.

Characteristics of the shot are:

(a) It happens in a same frame;
(b) The event is registered in a single pixel or adjacent pixels.

2.4.2 Description of the Algorithm

A lightning discharge will often illuminate some adjacent pixels during a period of time, which means the continuous response is produced in a specific area. Clustering algorithm of lightning events based on the relativity of time and space is adopted to convert events into groups and flashes, the random and abrupt flash is recognize as a shot [9]. As shown in Fig. 8(a), first, events are encoded into the Freeman chain code along the clockwise direction. Next, to form the code of an area, an arbitrary starting pixel (x, y) is chosen on boundary, the contour search takes place pixel-by-pixel around the boundary in an clockwise direction, the search stops and the chain code moves a step until the lightning event is detected [10]. Finally, the search completes until the chain code returns to the starting point at (x, y) or goes to the end. As shown in Fig. 8 (b), the chain code of the group consists of directions and positions, the direction of group is 3-4-5(group 1), 2-3-6-7(group 2), 4(group 3), 4-6-1(group 4), 2-2(group 7), 3-6-1(group 8), the chain code of group 5 and group 6 is only represented by the position of the pixel because they are one-pixel groups.

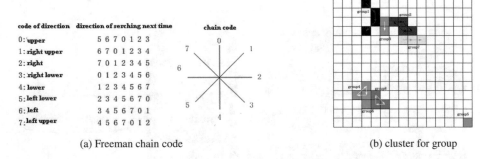

(a) Freeman chain code   (b) cluster for group

Fig. 8. Diagram of events cluster

Events in the single frame are first clustered into groups by the algorithm of freeman chain code. Groups are assigned to flashes base on the constraint of time and distance of flash, the flash composed with events including only in one frame should be filtered as shots.

The clustering process is shown in Fig. 8(b). There are twenty one events happened at different time with different color. Four events filled with black color are detected at the time of T_0, since events are simultaneous and register in adjacent pixels, they are collected into a group 1. The next time integration with red color data is assigned into a new group 2 at the time of T_1. At the time of T_2, five events filled with green color are converted into two groups 3 and 4. Then group 5 to 8 form sequential from time of T_3 to T_5. Group 1,2,3 and 6 are adjacent, group 6 occurred within T_f of group 3, these four groups are assigned to a flash A. Group 4,6 and 8 are assigned to a new flash B based on the same principle. Group 5 is composed of one event and far from the other groups, so it is recognized as a shot.

3   Experiment

In order to evaluate the performance of the lightning filter algorithm proposed in this paper, we selected a period from 10:06:00 UTC on March 29th to 00:06:00 UTC on March 30th, 2017, when a thunderstorm happened in the south and southwest China. The FY-4 satellite located at 99.5° east longitude at the same time. For the purpose of illustrating a comparison, the cloud-to-ground lightning strokes of the same period was acquired by the ground-based lightning network of China.

As shown in Fig. 9(a), the LMI detected 89468 events during one-hour period starting on 00:06:00(UTC) March 30th, as shown in Fig. 9(b), there are 22167 events left after filtering, only ∼24% of events as identified by the filter algorithm described in the text is true, with the rejection rate for this data set of ∼76%, the result is basically in accordance with the conclusion of LIS. Table 3 lists the percentage of events removed during each kind of the filtering process for the dataset. It is noticed that shots account for the most of events.

Table 3. Overview of false events filtering process

Filter	Total events	Removed by filter	% of original removed
Raw	89468	–	–
Ghost	89468	2	0.0022
Track	89466	225	0.2515
Shot	89241	67459	75.5919

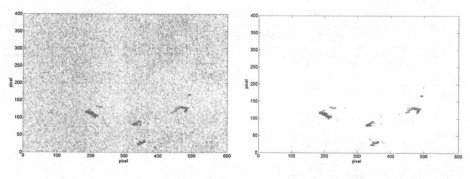

(a) events with all filters switched off   (b) events with all filters have been applied

Fig. 9. Results of filtering

The contrast result of LMI and ground-based net is shown by Fig. 10. Based on the result of geometric correction,the land mark of China is resampled to the field-of-view of LMI for the comparision. The detected intensity which refers to the detected events density at the satellite is represented by the colormap. We can find the position and the relative density of events in China are generally similar. This thunderstorm happened both in the southern Yunnan and southeastern Hunan. The difference of the detected intensity is mainly caused by two reasons, first, the detected target is different: the cloud-to-ground lightning can be detected from ground but both cloud and cloud-to-ground lightning can be detected from satellite, second, the detected intensity denotes the count of lightning strikes for ground but the count of events for satellite.

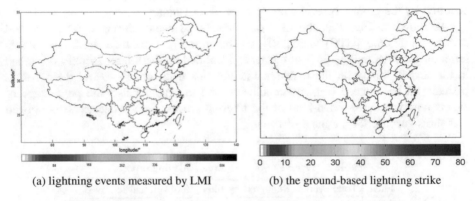

(a) lightning events measured by LMI   (b) the ground-based lightning strike

Fig. 10. Lightning monitoring results of LMI and ground based nets

Results of other one-hour periods are shown in Table 4. The average rejection rate reaches 77%.

Table 4. Performance of one-hour false events filtering process

Time (UTC)	Total events	Removed by filter	% of original removed	Time (UTC)	Total events	Removed by filter	% of original removed
10:06–11:06	130783	19996	84.7	17:36–18:06	59581	11746	80.3
11:06–12:06	155396	31520	79.7	18:06–19:06	106028	23780	77.6
12:06–13:06	149224	33237	77.7	19:06–20:06	135637	29233	78.4
13:06–14:06	150203	41907	72.1	20:06–21:06	161948	42128	74.0
14:06–15:06	128225	32916	74.3	21:06–22:06	160498	37935	76.4
15:06–16:06	112498	25653	77.2	22:06–23:06	154676	40683	73.7
16:06–17:06	133884	34301	74.4	23:06–24:00	150407	31267	79.2

The performance of short-term lightning tracing by LMI is evaluated on the data obtained on June 21th, 2017, when a serve storm happened in Beijing-Tianjin-Hebei region. FY-4 satellite located at 99.5° east longitude at the same time. As shown in Fig. 11(d), the color of the lightning detected from ground during 15:00–16:00 was green, it changed to be blue during 16:00–17:00 and red during 17:00–18:00, it was noticed that the lightning in this region moved from west to east during three hours. The data of LMI during the same period is processed by the proposed filtering method. The location and the movement of LMI lightning production as shown in Fig. 11(a)–(c) is accordance with the ground based monitoring result.

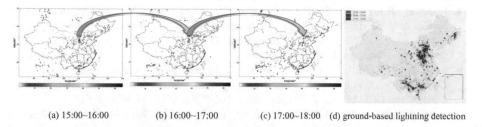

(a) 15:00~16:00   (b) 16:00~17:00   (c) 17:00~18:00   (d) ground-based lightning detection

Fig. 11. Tracing movement of lightning

4 Summary and Conclusions

The LMI of FY-4 satellite is the first geostationary lightning detection remote sensor of China, although the lighting cluster-filter algorithm has been proposed by NASA and also has been validated through LIS and GLM [11], but the onboard lightning detection and downloading principle of LMI is different from them, so the algorithm must be modified to make the process more efficient. In this paper, we have developed an improved lightning filtering algorithm based on the performance of LMI for minimizing the number of non-lightning events, which is used both at event level and at flash level. It has been proved that the experimental result is satisfactory for storm and flood warning and nowcasting with the monitoring region larger than ground based lightning detection.

In the future, we will dedicate our effort to four works as below:

(1) continually follow every kind of non-lightning events and the performance of the LMI;
(2) geometric correction for lightning location based on the height of cloud [12];
(3) calibration of production for obtaining clustering parameters by space-ground experiment;
(4) research on in-orbit radiometric calibration using deep convective clouds [13];
(5) short-term forecast model for rainfall by analyzing abundant in-orbit data [14].

It is foreseeable that the LMI on the FY-4 satellite will play more important role in the lightning nowcasting, lightning strike early warning and climate monitoring of China and surrounding regions.

References

1. Christian, H.J., Blakeslee, R.J., Goodman, S.J.: Lightning imaging sensor (LIS) for the earth observing system. NASA Technical Memorandum 4350 (1992)
2. Bao, S., Tang, S., Li, Y., Liang, H., Zhao, Y.: Real-time detection technology of instantaneous point-source multi-target lightning signal on the geostationary orbit. Infrared Laser Eng. **41**(9), 2930–2935 (2012)
3. Cao, D.: The development of product algorithm of the Fengyun-4 geostationary lightning mapping imager. Adv. Met. S&T **6**, 94–98 (2016)
4. Hui, W., Huang, F., Guo, Q.: Filtering of false signals in geostationary lightning detection by satellite. Meteorol. Sci. Technol. **43**, 805–813 (2015)
5. Buechler, D.E., Christian, H.J., Koshsk, W.J., Goodman, S.J.: Assessing the lifetime performance of the lightning imaging sensor (LIS): implications for the geostationary lightning mapper (GLM). In: XIV International Conference on Atmospheric Electricity, Rio de Janeiro, Brazil, pp. 1–4, 08–12 August 2011
6. Suszcynsky, D.M., Light, T.E., Davis, S., Green, J.L., Guillen, J.L.L., Myre, W.: Coordinated observations of optical lightning from space using the FORTE photodiode detector and CCD imager. J. Geophys. Res. **106**(D16), 17897–17906 (2001)
7. Daniels, J., Goldberg, M., Wolf, W., Zhou, L., Lowe, K.: GOES-R Algorithm Working Group (AWG). Atmospheric and Environmental Remote Sensing Data Processing and Utilization V: Readiness for GEOSS III, pp. 74560.1–74560.7 (2009)
8. Song, X., Yuan, S., Guo, H., Liu, J.: Pattern identification algorithm with adaptive threshold interval based extended Hough transform. Chin. J. Sci. Instrum. **35**(5), 1109–1117 (2014)
9. Goodman, S., Mach, D., Koshak, W., Blakeslee, R.: GLM lightning cluster-filter algorithm. NOAA Nesdis Center for Satellite Applications and Research Algorithm Theoretical Basis Document (2010)
10. Sun, J.: Contour representation and retrieval based on spatial feature and relativity of chain codes. J. Optoelectron. Laser **19**(8), 1112–1115 (2008)
11. Goodman, S., Blakeslee, R., Koshak, W., Mach, D.: High impact weather forecasts and warnings with the GOES-R geostationary lightning mapper (GLM). Marshall Space Flight Center (2011)
12. Chen, S.-B., Yang, Y., Cui, T.-F.: Study of the cloud effect on lightning detection by geostationary satellite. Chin. J. Geophys. **55**(3), 797–803 (2012)
13. Ma, S., Huang, Y.-X., Yan, W., Ai, W.-H., Zhao, X.-B.: Calibration of low-level light sensor using deep convective clouds. J. Infrared Millim. Waves **34**(5), 630–640 (2015)
14. Baker, M.B., Blyth, A.M., Christian, H.J., Latham, J., Miller, K.L., Gadian, A.M.: Relationships between lightning activity and various thoudercloud parameters: satellite and modeling studies. Atmos. Res. **51**(3–4), 221–236 (1999)

Land Use Type Change Detection by Landsat TM and Gaofen-1 Data - A Case Study at Jishui County of Jiangxi Province

Zhaopeng Zhang[1,2](✉), Zengyuan Li[1], Erxue Chen[1], and Xin Tian[1](✉)

[1] Institute of Forest Resources Information Technique,
Chinese Academy of Forestry, Yiheyuanhou, Beijing
People's Republic of China
zhaopengzhang1123@163.com, tianxin@caf.ac.cn
[2] College of Geomatics, Xi'an University of Science and Technology, Yanta
Road, Xi'an, People's Republic of China

Abstract. Due to natural and anthropic disturbance, the global land use types dramatically changed, especially during the past decades. Based on maximum likelihood classifier, this study applied the Landsat Thematic Mapper 5 (TM) and Chinese domestic high resolution satellite data, Gaofen-1, to investigate the changes of land cover types between 2008 and 2016 at Jishui county of Jiangxi province, the Central China. The validation using the forest field data showed that the overall classifications of 2008 and 2016 are 91.40%, 94.68%, respectively. Based on these two temporal classification maps, we quantitatively analyzed changes of each land use type and its transitions to others. The analysis showed that two types largely changed between the study years, the water area and forest area. The water area increased significantly, which accounted for 1.66% of the total area of the study area, and the total forest coverage rate increased 2.6% from 2008 (62.04%) to 2016 (64.64%).

Keywords: Land use type change · GaoFen-1 (GF-1)
Landsat Thematic Mapper 5 (TM)

1 Introduction

In the past decades, dramatic conversions among various land use patterns have occurred in China due to global climate change, population increase, urbanization, reform in regional land use policy, and etc. These changes have had major impacts on regional water cycles, biological diversity, terrestrial ecosystem productivity. In order to understand the current situation, trends, driving forces, and environmental impacts of land use changes, using multi-source and multi-temporal remote sensing data, such as, Landsat Thematic Mapper 5 (TM) [1], QuickBird [2], WorldView-2 [3], numerous studies have been performed. However, there are few research based on Chinese domestic high resolution satellite constellation data (i.e., Gaofen series data). The Chinese high resolution satellite constellation is merited by many key technologies, for example, it contains high spatial and temporal resolution, good attitude control

© Springer Nature Switzerland AG 2018
H. P. Urbach and Q. Yu (Eds.): ISSOIA 2017, SPPHY 209, pp. 173–179, 2018.
https://doi.org/10.1007/978-3-319-96707-3_19

technology with high precision and high stability, high reliability with 5–8 years of service life. Liu [5] and Hu [6] carried out land use classification and change detection based Gaofen-1 (GF-1) images over the Loess Plateau and Harbin city respectively. However, only several land cover types were considered, various forest types have not been discriminated in their studies.

The objective of this study is to apply the time-continuous and free Landsat Thematic Mapper 5 (TM) and Chinese GF-1 data to monitor and assess the land use type changes, especially for the forest types, over a typical county in the Central China.

2   Materials and Methods

2.1   Study Area and Data

Jishui county of Jiangxi province, was selected as the study area, as it was a typical changing county due to fast development over the past decade (Fig. 1). Jishui county is located in subtropical area, and the terrain consists of mountains, hills and plains, with the significant characteristics of hills. The county covers about 2509 km^2 (Fig. 2).

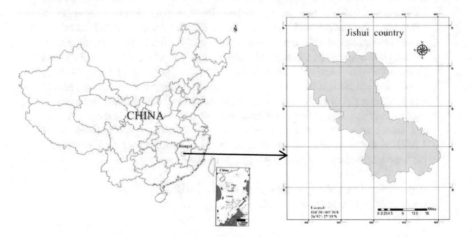

Fig. 1. The location of study area

A TM (acquired on July 25, 2008) and a GF-1 image (acquired on July 29, 2016) covering Jishui county were obtained from United States Geological Survey (USGS) website (http://glovis.usgs.gov/) and China Center for Resources Satellite Data and Application. GF-1 is a kind of Chinese high resolution satellite constellation data, with 16 m resolution for multi-spectral image 8 m panchromatic image.

2.2   Change Detection

The specific technical flowchart of this study is shown in Fig. 3.

In this study, data pre-processing, including radiometric calibration, atmospheric correction and geometric correction, was conducted by use of ENVI5.4.1 software.

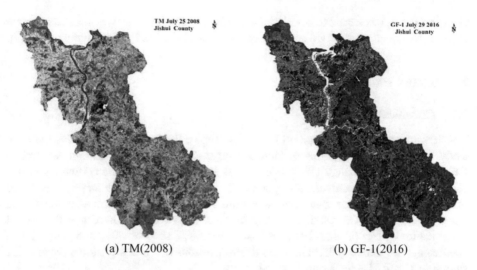

(a) TM(2008)   (b) GF-1(2016)

Fig. 2. TM (Red: band 3; Green: band 2; Blue: band 1) and GF-1 (red: band 4; Green: band 3; Blue: band 2) images of Jishui county.

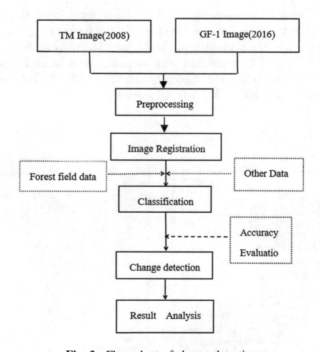

Fig. 3. Flow chart of change detection

In particular, the GF-1 image was co-registered with Landsat TM images with overall error less than 1 pixel (30 m). The land use type samples were collected from the field survey, which were used as the training and testing data. The maximum likelihood classifier (MLC) was used to discriminate the local main land use types. Land use type changes were analyzed by two temporal classification maps and their statistic results [7].

3   Results and Analysis

3.1   Classification Results

Considering the present situation of land use types, the field work samples over the study area, and the categories of land use types proposed by the Intergovernmental Panel on Climate Change (IPCC), the local land cover types were divided into 9 classes, including water/wetlands, grassland, cultivated land, settlements, forest and others, where the forest was subdivided into coniferous forest, broad-leaved forest, bamboo and shrub. The validation using the field survey data showed that the overall classifications of 2008 and 2016 are 91.40, 94.68%, respectively, and both Kappa coefficients are above 0.88. The classification results and classification accuracy are shown in Fig. 4, Tables 1 and 2, respectively.

3.2   Analysis of Land Cover Type Changes

As shown in Tables 3 and 4, changes of land cover types between 2008 and 2016 mainly occurred in water, settlements forest land; other types changed a little. Among forestland, the changes among coniferous forest, broad-leaved forest and bamboo were obvious. In details, 39.2% of bamboo were converted into broad-leaved forest. The conversion of cultivated land into bamboo forest accounted for 11.76%; and 19.35% of coniferous forest were converted into broad leaved forest.

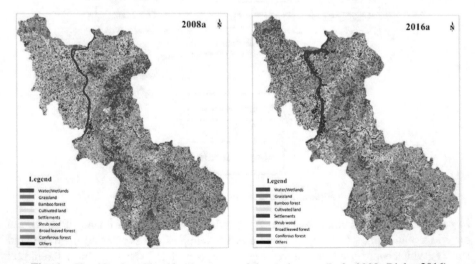

Fig. 4.   Classification Classification maps of the study area (Left: 2008; Right: 2016)

Table 1. The confusion matrix of classification accuracy in 2008 (Pixels)

2008a	Coniferous	Broad - leaved	Bamboo	Grassland	Water	Cultivated	Settlement	Shrubbery	Total
Coniferous	520	14	0	25	1	0	0	32	592
Broad-leaved	12	301	5	46	0	0	0	32	396
Bamboo	0	64	596	14	0	4	0	19	697
Grassland	1	0	0	9	9	2	3	0	24
Water	0	0	0	51	1943	0	0	0	1994
Cultivated	26	0	1	25	8	1820	3	67	1950
Settlement	4	0	0	4	11	4	193	14	230
Shrubbery	0	0	2	0	4	0	0	5	11
Total	563	379	604	174	1976	1830	199	169	5894

Overall Accuracy = (5387/5894) = 91.3980% Kappa Coefficient = 0.8863.

Table 2. The confusion matrix of classification accuracy in 2016 (pixels)

2016a	Coniferous	Broad - leaved	Bamboo	Grassland	Water	Cultivated	Settlement	Shrubbery	Total
Coniferous	1058	3	0	0	0	0	0	0	1061
Broad-leaved	49	623	11	0	0	0	0	7	690
Bamboo	0	113	1264	0	0	0	0	0	1377
Grassland	0	0	0	321	6	0	2	0	329
Water	0	0	0	0	1443	0	0	0	1443
Cultivated	51	0	0	3	6	1800	2	0	1862
Settlement	7	0	0	0	46	2	380	0	435
Shrubbery	7	0	0	1	0	17	0	300	325
Total	1172	739	1275	325	1501	1819	384	307	7522

Overall Accuracy = (7189/7522) = 94.6823% Kappa Coefficient = 0.9473.

Table 3. The percentage of various types of land use to the total area (%)

	Grassland	Water	Coniferous	Broad-leaved	Cultivated	Settlement	Shrubbery	Bamboo	Total
2008a	0.965	2.717	13.499	21.259	28.619	5.729	5.744	21.540	100
2016a	3.122	4.375	22.501	22.585	22.592	5.270	4.033	15.621	100

Generally, the area of settlements almost unchanged during 2008 and 2016 over the study area. The forest coverage increased by 2.6% points from 62.04% (2008) to 64.64% (2016).

Table 4. Transform situation matrix of land cover (%)

2008–2016	BG	GL	WW	CF	BL	CL	SM	BF	SB
UC	0.000	0.000	0.000	0.000	0.000	0.000	0.000	0.000	0.000
BG	99.917	0.041	0.366	0.160	0.159	0.065	0.042	0.145	0.113
GL	0.002	12.581	10.853	1.775	0.920	4.593	10.419	0.974	2.757
WW	0.001	18.077	70.456	1.991	0.944	3.032	5.715	1.1970	6.607
CF	0.022	8.029	1.330	54.566	37.903	11.966	4.769	14.341	2.790
BL	0.027	5.328	0.765	19.347	34.248	11.653	4.443	39.199	8.993
CL	0.011	25.779	4.579	6.584	7.644	43.100	41.424	11.762	42.984
SM	0.001	19.368	9.298	1.464	1.004	8.251	18.938	1.466	11.383
SB	0.005	7.000	1.731	4.791	2.864	5.973	8.411	1.259	1.801
BF	0.013	3.796	0.622	9.322	14.314	11.366	5.840	29.657	22.572

Notes: WW: Water/Wetlands; GL: Grassland; CL: Cultivated land; SM: Settlements; BL: Broad-leaved forest; BF: Bamboo forest; BG: Background; UC: Unclassified; CF: Coniferous forest; SB: Shrubbery.

4 Conclusion

This study investigated the land cover changes over a typical developing county by use of two temporal remotely sensed images. Under the premise of the overall classification accuracy of 2008 and 2016 were 91.40%, 94.68%, respectively, the changes detections were reliable. The result showed that, due to the fast development, urbanization, and reform in land use policy, the local land use types dramatically changed. In general, the cultivated areas largely decreased by 6.03%, but the grassland and forest area increased by 2.16%, and 2.60%, respectively, which can be ascribed to the conduction of the "Grain for Green" project. In particular, broad-leaved forest and coniferous forest increased by 1.33%, 9.00%, respectively. Bamboo forest and shrubbery decreased by 5.92, and 1.71%, respectively, from 2008 to 2016. Moreover, water area significant increased mainly due to the construction of the reservoir.

Acknowledgements. We thank to China Center for Resources Satellite Data and Application and United States Geological Survey for providing the GF-1 image and the TM images with excellent quality. The authors also would like to thank Jiangxi Forest Inventory and Planning Institute and Jishui County Land Resources Bureau for providing some of the ground true data and the second national land survey land classification data for the validations. This word was funded by the Fundamental Research Funds for the Central Non-profit Research Institution of CAF under Grant CAFYBB2017QC005 and the National Natural Science Foundation of China (21-Y30B05-9001-13/15-1).

References

1. Li, J.C., Liu, H.X., Liu, Y., Su, Z.Z., Du, Z.Q.: Land use and land cover change processes in China's eastern Loess Plateau. Sci. Cold Arid Reg. **7**(6), 722–729 (2015)
2. Chen, S.J., Shan, D.D., Zhao, W.C.: Comparison and analyses on land cover classification methods using remote sensing. J. Liaoning Tech. Univ. (Nat. Sci.) **29**(4), 567–570 (2010)
3. Lu, D.S., Hetrick, S., Moran, E., Li, G.Y.: Detection of urban expansion in an urban–rural landscape with multitemporal QuickBird images. J. Appl. Remote Sens. **4**(10), 201–210 (2010)
4. Wu, H., Cheng, Z.P., Shi, W.Z., Miao, Z.L., Xu, C.C.: An object-based image analysis for building seismic vulnerability assessment using high-resolution remote sensing imagery. Nat. Hazards **71**(1), 151–174 (2014)
5. Liu, L.W.: Dynamic changes of land cover based on GF-1 and TM in loess hilly region—the case of Taigu county. Sci. Technol. Eng. **16**(07), 122–128 (2016)
6. Hui, Q.W., Zuo, J.J., Liu, J., Liu, D.D., Wu, C.C., Zhang, Y.J.: Research on GF-1 in the application of the land cover dynamic monitoring. Geomat. Spat. Inf. Technol. **36**(6), 63–66 (2016)
7. Zhang, Z.P., Li, Z.Y., Li, Q.H., Hao, R.X.: Dynamic analysis on the vegetation coverage changes of Minqin Oasis based on GF-1 from 2013 to 2015. J. Southwest For. Univ. **37**(2), 176–183 (2017)

The Dark-Signal Real-Time Correction Method of CCD Digital Image

Chunmei Li[✉], Yuanrong Guo, Feiran Zhang, Fang Dong,
Yuting Hu, and Shuli Dong

Beijing Institute of Space Mechanics and Electricity,
Key Laboratory for Advanced Optical Remote Sensing Technology
of Beijing, Beijing 100094, China
15910620602@126.com

Abstract. Dark signal is the non-negligible component of the CCD (Charge Coupled Devices) camera noise. The most straightforward way to eliminate dark signal is cooling the CCD chip. Moreover, the dark-signal image correction technique gets the more extensive application because of its lower development costs and better dark-signal correction effect. In this article, two methods for estimating the average of dark signals based on temperature and dark pixels are introduced, as well as the method for estimating the pixel nonuniformity of dark signal and for pixel-level dark signal correction. Experiment results show that the method in this article can achieve higher accuracy in dark-signal correction.

Keywords: Dark signal · CCD · Real-time correction

1 Introduction

Dark signal is produced by camera when the CCD (Charge Coupled Devices) is in complete darkness, as in [1]. The main component is Dark current. Dark current is intrinsic to semiconductors and naturally occurs through the thermal generation of minority carriers. The level of dark current generated determines the amount of time a potential well can exist to collect useful signal charge. This time is not very long, therefore CCD users must deal with the dark current problem head on as in [1–3].

The dark-current noise is intimately related to the temperature, for example, the dark current increases twice while every 7 or 10° increase in 0–50 °C temperature, as in [4]. In a weak signal condition, CCD is observed by a long-time-integral method, which leads to that dark current is the main factor reducing the dynamic range of imaging device, as in [5, 6]. The most direct way to eliminate dark signal is cooling the CCD chip. However the dark-signal image correction technique has the more extensive application for its lower development costs and better dark-signal correction effect. Therefore, the technology separating dark signal from CCD digital image data is the dark-signal correction method of image, as in [7].

In dark-signal real-time correction method of CCD digital image, CCD dark pixels, device temperature and other information are used as the dark-signal estimation, and the dark-signal pixel nonuniformity correction is implemented at the same time. In the end, the effect of the equivalent gain of the imaging circuit is also considered to achieve

© Springer Nature Switzerland AG 2018
H. P. Urbach and Q. Yu (Eds.): ISSOIA 2017, SPPHY 209, pp. 180–188, 2018.
https://doi.org/10.1007/978-3-319-96707-3_20

accurate correction of the dark signal in the image. This method reduces the dark-signal level in the image greatly and eliminates the fixed graphic noise in dark signal. All in all, this technique is highly utilized in the high-precision CCD imaging applications because of its better real-time and higher accuracy.

2 Dark-Current Characteristics Analysis

2.1 Dark Current Analysis

Surface and depletion dark current, the two main contributors to dark current, exhibit the same temperature dependence. Therefore, we can use (1) as the general dark current formula for the CCD, as in [4, 8].

$$D_R(e^-) = CT^{1.5}e^{-E_g/2kT} \tag{1}$$

Where $D_R(e^-)$ is the average dark current generated (e$^-$/s/pixel), C is constant, T is operating temperature (K), E_g is silicon band-gap energy, k is Boltzmann's constant $(8.62 \times 10^{-5}$ eV/K).

The bandgap energy varies with operating temperature following the empirical formula (2), as in [8].

$$E_g = 1.1557 - 7.021 \times 10^{-4}T^2/1108 + T \tag{2}$$

The constant C can be solved at the room temperature (300 K, 27 °C), yielding (3), as in [8].

$$C = \frac{D_{FM}P_S}{qT_{RM}^{1.5}e^{-E_g/(2kT_{RM})}} \tag{3}$$

Where P_S is pixel area (cm^{-2}), T_{RM} is room temperature (300 K), D_{FM} is called the "dark current figure of merit" at 300 K (nA/cm^2). Substituting (3) into (1) produces the final dark current formula (4), as in [8, 9].

$$D_R(e^-) = 2.5 \times 10^{15}P_S D_{FM}T^{1.5}e^{-E_g/2kT} \tag{4}$$

For instance, these curves, given in Fig. 1 are about changes of dark-current response and temperature with 10 pA/cm^2 quality index in 12-μm pixel-size CCD.

2.2 Distribution Characteristics of Dark-Current Noise

The dark current noise is the important component of the dark current output signal, which refers to dark-current shot noise and dark-current nonuniformity.

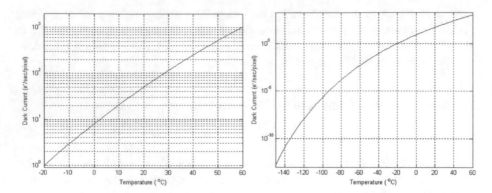

Fig. 1. Relation between dark-signal response and temperature

The dark-current shot noise is governed by Poisson's statistics, its basic equation is (5).

$$N_{DSN} = D(e^-)^{1/2} \tag{5}$$

Where N_{DSN} is the dark shot noise (*rms e^-*), and $D(e^-)$ is the average dark current level ($e^-/pixel$).

Dark-current nonuniformity is the variation of dark current from pixel to pixel. The noise that is produced is (6).

$$N_{DN}(e^-) = \sigma_{DN}D(e^-) \tag{6}$$

where N_{DN} is dark current nonuniformity noise (*rms e^-*), σ_{DN} is dark current nonuniformity factor.

The total noise generated by dark current is the quadrature sum of dark current shot noise and dark current nonuniformity, which is (7).

$$N_{DC}(e^-) = \left[(\sigma_{DN}D(e^-))^2 + D(e^-) \right]^{1/2} \tag{7}$$

When the average dark current is low, dark shot noise prevails. As dark current increases, nonuniformity begins to dominate. We can find the maximum integration time before the dark current nonuniformity equal the shot noise.

Therefore, the dark current noise is shown in Fig. 2.

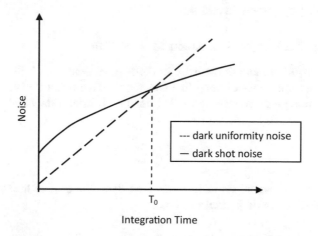

Fig. 2. Curves about noise distribution and integration time

3 Dark Signal Correction

3.1 Average Dark-Signal Estimation Method One

Known from the dark current formulas, for a certain CCD, its pixel size and quality factor are determined. Dark signal varies with operating temperature, integration time, circuit equivalent gain and noise. It can be shown.

$$D_R = f(T, t, G, N_d) = 2.5 \times 10^5 \cdot P_S D_{FM} T^{1.5} e^{-E_g/2kT} \cdot t \cdot A + B \qquad (8)$$

Where t is integration time, A is circuit equivalent gain, B is additive noise produced by circuit. Merging constants P_S, D_{FM}, and A in (8), the new formula can be shown as follow (9).

$$D_R = r \cdot T^{1.5} e^{-E_g/2kT} \cdot t \cdot A + B \qquad (9)$$

Where r is constant factor determined by pixel size and quality factor as well as some other terms.

When the operating temperature variety is less than 30 °C, achieving dark-signal estimation by using function fitting method is highly accuracy and can be easily implemented in hardware. It follows the empirical formula (10).

$$D_{RE,T} = (C_1 \cdot T^3 + C_2 \cdot T^2 + C_3 \cdot T + C_4) \cdot t \cdot A + B \qquad (10)$$

Where $D_{RE,T}$ is the dark signal estimation based on temperature, C_1, C_2, C_3, C_4 are coefficients.

In above formula, coefficients C_1, C_2, C_3, C_4 and B are determined by camera calibration experiment, besides, the temperature T, integration time t, as well as circuit gain A provided by camera in real time.

3.2 Average Dark-Signal Estimation Method Two

Some CCD provide a more direct way for dark-signal detecting, which is dark pixel. When there are dark pixels existing in CCD, the average of dark pixels is commonly used as estimation value of dark signal. This method is brief, direct and high precision.

$$D_{RE,P} = \frac{1}{M} \sum_{i=1}^{M} d(i) \tag{11}$$

Where $D_{RE,P}$ is dark signal estimation based on dark pixel, M is the total number of dark pixels, $d(i)$ is i-th dark pixel response.

3.3 Estimation of Dark-Signal Nonuniformity

Dark current shot noise can only be eliminated by cooling the CCD. But Dark current nonuniformity also removed by image processing techniques. Because its distribution is relatively stable and can be estimated by image processing. Equation (11) is the dark signal nonuniformity, as in [2].

$$N_{DN}(n) = \overline{D_R(n)} - \overline{D_R} \tag{12}$$

Where $\overline{D_R(n)}$ is average dark signal of n-th pixel, $\overline{D_R}$ is average dark signal of all pixel, $N_{DN}(n)$ is dark signal nonuniformity.

When CCD is exposed in light, the dark signal of each pixel cannot be measured, which leads to an indirect method to obtain the dark-signal nonuniformity noise. A linear dark signal nonuniformity estimation based on dark pixels is (13).

$$N_{DN}(n) = k_n \cdot D_{RE} + b_n \tag{13}$$

Where D_{RE} is average dark signal estimation, k_n, b_n are coefficients.

Therefore, average dark signal and dark-signal nonuniformity form the dark signal estimation, as (14).

$$D_{RE}(n) = D_{RE} + N_{DN}(n) = k_n' \cdot D_{RE} + b_n \tag{14}$$

Where k_n', b_n are coefficients, $D_{RE}(n)$ is dark signal estimation of n-th pixel.

3.4 Dark-Signal Correction

The dark-signal correction is truly the image information denoising method based on dark signal. In the output image of CCD, dark signal is useless information, its existing reduces the image quality. Besides, the real purpose of dark-signal correction is to

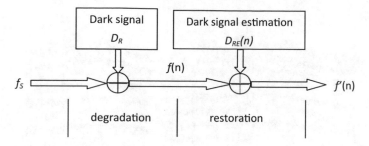

Fig. 3. Dark signal correction sketch map

eliminate or decrease the effect of dark signal in image information, and increase image quality as well. Figure 3 shows the sketch map of dark signal correction.

The mathematical expression of correction module is (15):

$$f'(n) = f(n) - D_{RE}(n) \tag{15}$$

Where $f(n)$ is n-th pixel signal before dark signal correction, $D_{RE}(n)$ is dark signal estimation of n-th pixel, $f'(n)$ is n-th pixel signal after dark signal correction.

The CCD dark signal correction based on dark pixel follows (16):

$$f'(n) = f(n) - k'_n \cdot D_{RE,P} - b_n \tag{16}$$

Where k'_n and b_n are nonuniformity correction coefficients of n-th pixel.

The CCD dark-signal correction based on temperature follows (17):

$$f'(n) = f(n) - k'_n \cdot D_{RE,T} - b_n \tag{17}$$

The former is high accuracy, but its disadvantages are obvious, such as abundant calculation, low speed and consuming more resources. However, the latter is much easier to achieve, faster to calculate, and less computation and resources needed.

4 Experiment and Results

In a experiment, the CCD has 3072 pixels, 2 output channels, several dark pixels distributed on each channel. Device has 14-bit quantized output. There are 24 sets of original dark signal data and dark pixels data, corresponding temperature data are obtained at the same time. The original dark signal distribution is shown in Fig. 4. These data are separately processed by the two method above. The results of dark signal correction using method 1 are shown in Fig. 5, while the results of the other method is shown in Fig. 6.

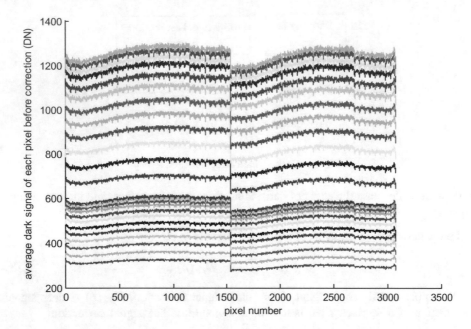

Fig. 4. The average of original dark signal of each pixel

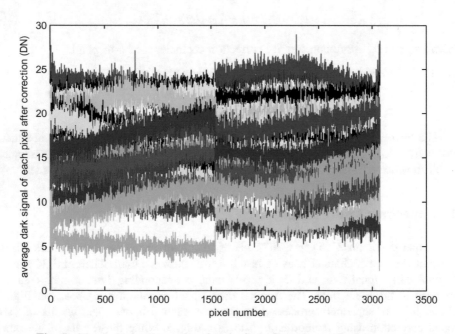

Fig. 5. The dark signal average of each pixel after using method 1 correction

The dark signal average and the dark signal nonuniformity are the most

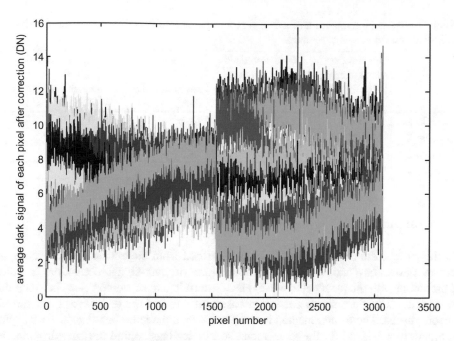

Fig. 6. The dark signal average of each pixel after using method 2 correction

comprehensive and representative indicators. These two indicators of before and after correction are compared accurately. The results using method 1 are shown in Table 1, the results using method 2 are shown in Table 2 as well.

Table 1. The comparing result of dark signal average and the dark signal nonuniformity before and after correction using method 1

	Original dark signal average (DN)	Corrected dark signal average (DN)	Original dark signal nonuniformity (DN)	Corrected dark signal nonuniformity (DN)
Sample 1	305.80	10.84	14.76	1.31
Sample 2	676.22	22.76	18.77	1.67
Sample 3	951.16	12.48	23.24	1.33
Sample 4	1251.3	11.06	29.54	1.61

Table 2. The comparing result of dark signal average and the dark signal nonuniformity before and after correction using method 2

	Original dark signal average (DN)	Corrected dark signal average (DN)	Original dark signal nonuniformity (DN)	Corrected dark signal nonuniformity (DN)
Sample 1	305.80	7.63	14.76	0.95
Sample 2	676.22	9.31	18.77	1.82
Sample 3	951.16	8.25	23.24	2.20
Sample 4	1251.3	5.39	29.54	1.96

5   Conclusion

In this article, the dark signal of CCD is described from theoretical analysis to application derivation, and a mathematical module of dark-signal correction is built. Combining with the practical applications of a certain circuit, comparisons between the two methods of dark signal correction based on temperature and dark pixels can be accomplished. The former method is less precise than the latter because the dark signal estimation is delayed for the temperature sensitivity lags behind the nuclear temperature since only the surface temperature of CCD devices can be obtained during the experiments. All in all, the dark signal correction method of digital image mentioned in this article can get high-precision results and provide a feasible dark signal real-time correction method for CCD cameras, thus improving the image quality thereby.

References

1. Janesick, J.R., Elliott, T., Collins, S., Blouke, M.M., Freeman, J.: Scientific charge-coupled devices. Opt. Eng. **26**, 692–714 (1987)
2. Pan, C., Weng, F., Beck, T., Liang, D., Devaliere, E., Chen, W., Ding, S.: Analysis of OMPS in-flight CCD dark current degradation. In: IGARSS, pp. 1966–1969 (2016)
3. Li, Y.F., Li, M., Si, G., Guo, Y.: Noise analyzing and processing of TDI-CCD image sensor. Opt. Precis. Eng. **15**, 1196–1202 (2007)
4. Chen, J., Zhang, Y., Liu, Y.: A CCD dark current auto-eliminating method. Sci. Technol. Eng. **15**, 15–19 (2015)
5. Shu, P.: CCD pixel model and simulation. University of Electronic Science and Technology of China (2009)
6. Shang, Y., Zhang, J., Guan, Y., Zhang, W., Pan, W., Liu, H.: Design and evaluation of a high-performance charge coupled device camera for astronomical imaging. Meas. Sci. Technol. **20**, 104002–104011 (2009)
7. Ma, B., Shang, Z., Hu, Y., Liu, Q., Wang, L., Wei, P.: A new method of CCD dark current correction via extracting the dark information from scientific images. Proc. SPIE **9154**, 91541T (2014)
8. Janesick, J.R.: Scientific Charge-Coupled Device. SPIE Press, Bellingham (2001)
9. Su, L.: The simulation model of CCD Image Sensors. University of Electronic Science and Technology of China (2014)

Study on Modeling and Simulation of LEO Satellite's Optical Reflection Characteristic Based on Davies Bidirectional Reflection Distribution Function

Fan Bu$^{(\boxtimes)}$, Yuehong Qiu, and Dalei Yao

Space Optics Laboratory, Xi'an Institute of Optics and Precision Mechanics,
CAS, Xi'an 710119, Shaanxi, China
bufan1986@opt.ac.cn

Abstract. Based on background radiation, material properties, orbit parameters, and the relative motion between LEO satellite body and solar panel, Davies-BRDF (bidirectional reflection distribution function) was used to establish the mathematical models of spectral and imaging characteristic of LEO satellite and calculate the satellite irradiance on entrance pupil and image plane of detector. Here, the surface of LEO satellite was divided into areas and each area was divided into grids. According to the orbit parameters of LEO satellite and detector, imaging condition of LEO satellite was analyzed and the judgment basis of imaging angle and scale was given. Taking different materials and shapes of satellites for example, these input parameters (the properties of structure, materials and orbit) were used to calculate the reflected radiation of every grid of satellite surface, and establish the mathematical models of reflection characteristic. The body coordinate of satellite was set and the relative positions of satellite, background radiation and detector were determined making use of coordinate conversion. In this paper, the imaging degrading effects were taken into consideration in the process of simulation. Simulated results of the satellite optical characteristic were decided by the relative position of the solar, the earth and the LEO satellite, as well as its geometrical shapes, flight figure and surface materials. The imaging results demonstrate that Davies-BRDF can simulate the reflection characteristics of the LEO satellite precisely and the simulated images was more accurate.

Keywords: BRDF · Davies · LEO satellite · OpenGL · Imaging simulation

1 Introduction

With the development of space, materials and information technology, space technology is also developing rapidly. A total of more than 6600 satellites have been launched around the world which are used in meteorological, navigation, communications, earth resources, geodesics, science and technology, etc. The development of space technology is not only the embodiment of a comprehensive national power, but also plays a vital role on the country's economy, scientific research, national defense, social stability.

© Springer Nature Switzerland AG 2018
H. P. Urbach and Q. Yu (Eds.): ISSOIA 2017, SPPHY 209, pp. 189–198, 2018.
https://doi.org/10.1007/978-3-319-96707-3_21

In the process of developing space technology, the space exploration system can detect the space environment at a more close distance and accurately monitor the operating status of the LEO satellite. Besides, this system can improve the efficiency of space utilization, and increase the breadth and depth of space exploration. For example, with the rapid development of human exploration of space, more and more space wastes affect the space exploration activities seriously, such as rocket wreckage, space debris. In order to ensure the safe operation of space resources, these work needs the support from the space exploration system. Therefore, the world countries regard the space exploration system as the focus of space technology development [1, 2].

To use the optical equipment to detect the LEO satellite, the primary task is to analyze the optical characteristic of the target. Because this kind of target does not light, the main analysis of the target is the optical characteristic of surface reflection. Many scholars at home and abroad have carried out this research based on geometrical optics and radiation theory of optical characteristic analysis. In this paper, the descriptive function of the target surface optical properties-Davies bidirectional reflection distribution function is used to analyze the visible light scattering characteristic of the target. The visible light scattering characteristic analysis based on the non-spontaneous optical LEO satellite is playing a great significance in the detection and recognition of LEO satellite [3, 4].

2   The Definition of BRDF

The Bidirectional Reflection Distribution Function (BRDF) was presented by American scholar Nicodemus in 1970. It shows the reflection characteristic of the surface at any angle of view under different incident angles, as shown in Fig. 1. BRDF is an important function to describe the diffuse reflection properties of the material. It is the ratio of the radiance to the incident irradiance of the optical radiation [5, 6]. The mathematical expression is given by

$$f_r(\theta_i, \varphi_i; \theta_r, \varphi_r) = \frac{dL_r(\theta_i, \varphi_i; \theta_r, \varphi_r)}{dE_i(\theta_i, \varphi_i)} \tag{1}$$

where θ_i is the incident angle of the incident light, φ_i is the azimuth angle of the incident light, θ_r is the radiance of the reflected light, and φ_r is the azimuth angle. L_r is the radiance of the element dA along the direction (θ_r, φ_r), and the unit is $W/m^2 \cdot sr \cdot \mu m$. E_i is the surface irradiance of the incident light along the direction (θ_i, φ_i), and the unit is $W/m^2 \cdot \mu m$. f_r is the ratio of the radiance emitted along the direction (θ_r, φ_r) to the irradiance produced by the measured surface along the direction (θ_i, φ_i). BRDF is the function of incident angle (θ_i, φ_i), reflection angle (θ_r, φ_r), and wavelength λ. It is determined by the target surface roughness, dielectric constant, irradiation wavelength, polarization and other factors. From the above, it can be seen that BRDF of the target surface can correlate the background irradiance incident on the target surface with the diffuse reflectance from the target to the background radiation. Moreover, BRDF can be used to analyze the light reflection characteristic of the target.

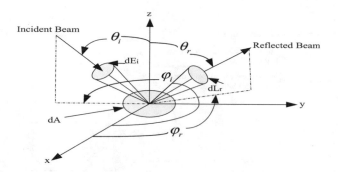

Fig. 1. The definition of the related parameters of BRDF

There are many models of BRDF. According to the surface condition of LEO satellite, this paper chooses Davies model which is based on the actual rough surface micro-geometric structure of the statistical characteristics of the introduction of mathematical formulas. The model assumes that the target surface consists of small randomly distributed faces [7]. Based on the Kirchhoff approximation, the expression is described as:

$$f_r(\theta_i, \varphi_i; \theta_r, \varphi_r) = \frac{\rho}{\cos \theta_i \cos \theta_r} \pi^3 \left(\frac{a}{\lambda}\right)^2 \left(\frac{\sigma}{\lambda}\right)^2 (\cos \theta_i + \cos \theta_r)^4$$
$$\times \exp\left\{-\left(\frac{\pi a}{\lambda}\right)\left[\sin^2 \theta_i + \sin^2 \theta_r + 2\sin \theta_i \sin \theta_r \cos(\varphi_i - \varphi_r)\right]\right\} \tag{2}$$

where θ_i is the incident zenith angle, θ_r is the zenith angle of the observation, φ_i is the incident azimuth, and φ_r is the observation azimuth. ρ is the reflectivity, σ is the surface roughness RMS, and a is the surface autocorrelation length.

In summary, if the LEO satellite surface is divided into several faces, the mathematical expression of BRDF can be introduced by

$$dL_r(\theta_i, \varphi_i; \theta_r, \varphi_r) = f_r(\theta_i, \varphi_i; \theta_r, \varphi_r) dE_i(\theta_i, \varphi_i) \tag{3}$$

The light reflection of any surface elements is obtained from the above equation, and then the light reflection distribution of the target is obtained using the integral of all facets. According to the above analysis, the BRDF model is used to analyze the reflection characteristics of the spatial target in the visible light band.

2.1 Modeling for the Visible Light Scattering Characteristics of the LEO Satellite

For deep LEO satellites, the visible background radiation mainly includes direct sunlight (sunlight), Earth and its atmospheric reflection radiation (Earthlight) and other stars of the reflection of radiation (Starlight) [8, 9] (Fig. 2).

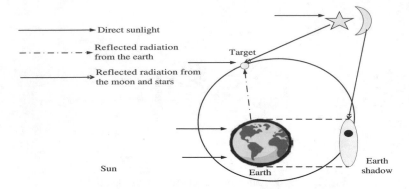

Fig. 2. Sketch of target background radiation

In order to accurately describe the reflection characteristics of the target surface, the surface of the target surface is divided according to the characteristics of the surface material. As shown in Fig. 3, each region is meshed. Due to the ups and downs of the cladding material of the satellite surface, each partition is divided into small plane. For each face, BRDF is used to establish its scattering characteristic model. Finally, all the components of the scattering element superimposed, and the result is the entire LEO satellite scattering characteristics.

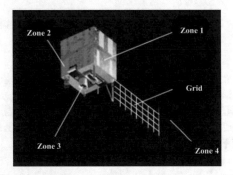

Fig. 3. Sketch of mesh and decomposition of the target

2.2 The Computation of the Direct Sunlight

According to the Planck blackbody radiation formula, the radiation intensity of the sun is shown in Eq. (4), and the irradiance from the sun to the target surface is shown in Eq.

$$M = \int_{\lambda_1}^{\lambda_2} \frac{c_1}{\lambda^5 [\exp(c_2/\lambda T) - 1]} d\lambda \tag{4}$$

$$E_{sun} = \frac{R_{sun}^2 M}{A_u^2} \tag{5}$$

c_1 is the first blackbody radiation constant, and the value is 3.741844×10^8 (W/m$^2 \cdot$ μm^4). c_2 is the second blackbody radiation constant, and the value is 14,388 (μm $\cdot$ K). T is the average temperature of the sun, and the value is 5900 K. R_{sun} is the radius of the solar, and the value is 6.9599×10^8 m. Au is the mean daily ground distance, and the value is $1.49597892 \times 10^{11}$ m.

2.3   The Computation of the Reflected Radiation from the Earth

Assuming that the earth is a diffuse reflector, the reflection of the solar radiation follows the Lambert law and is uniform throughout. And the reflection spectrum is similar to the solar spectrum. The reflectivity generally takes the average reflectance 0.35 of the earth. Therefore, the earth's reflection can be expressed as

$$E_{earth} = \rho E_{sun} \tag{6}$$

2.4   Modeling of Target Reflection Characteristics

(1) Modeling the reflection characteristics of direct sunlight

Since the target is far from the sun, the sun can be considered to be parallel to the target. According to the definition of BRDF, the radiance of the dots by solar radiation is given by the following equation

$$L_p(\lambda, \theta_i) = f_r E_{sun} \cos \theta_i \tag{7}$$

where $L_p(\lambda, \theta_i)$ is the spectral radiance value of the corresponding wavelength of the target surface, and the unit is W/m$^2 \cdot$ sr $\cdot$ μm. E_{sun} is the solar irradiance. θ_i is the angle between the normal to the target microbata dA and the sun Direction. f_r is BRDF of the surface dA of the satellite.

The spectral intensity of the spectral radiation generated by the surface dA is

$$dI_\theta = L_p(\lambda, \theta_i)dA = f_r E_{sun} \cos \theta_i dA \tag{8}$$

Assuming that the surface of the target is similar to Lambert surface, the spectral luminous intensity in the observation direction of the angle θ_r is defined as

$$dI_{\theta_r} = L_p(\lambda, \theta_i)dA = f_r E_{sun} \cos \theta_i \cos \theta_r dA \tag{9}$$

If the distance between the target and the detector is R, the spectral radiation of the satellite surface microbata dA reflecting the solar radiation at the detector into the pupil is expressed as

$$E_{dA} = \frac{dI_{\theta_r}}{R^2} = \frac{f_r E_{sun} \cos \theta_i \cos \theta_r dA}{R^2} \tag{10}$$

(2) Modeling of the reflection properties of the Earth and the atmosphere

According to the formula, the spectral radiation of the satellite target surface dA reflecting the Earth and the atmospheric light at the detector into the pupil of is

$$E'_{dA} = \frac{f'_r E_{earth} \cos \theta'_i \cos \theta'_r dA}{R^2} = \frac{f'_r \rho E_{sun} \cos \theta'_i \cos \theta'_r dA}{R^2} \tag{11}$$

In the Eq, θ'_i is the angle between the normal of the surface dA of the target and atmosphere reflected in the direction of the sun, and φ'_r is the angle between the normal of the surface dA of the target and detector observation direction.

In the visible band λ_1–λ_2 (0.38–0.76 μm), Eq. (12) is used to carry out the integral of the visible surface. Then the irradiance of the satellite reflecting the surface of the sun direct light, the Earth and the atmospheric reflecting light in the detector into the pupil can be expressed as

$$E = \iint E_{sun}(\lambda) \frac{f_r(\theta_i, \varphi_i, \theta_r, \varphi_r, a, e, d) E_{sun} \cos \theta_i \cos \theta_r \cos \theta_d dA}{R^2} d\lambda dA$$
$$+ \iint E_{sun}(\lambda) \frac{\rho f'_r(\theta'_i, \varphi'_i, \theta'_r, \varphi'_r, a, e, d) \cos \theta'_i \cos \theta'_r \cos \theta'_d}{R^2} d\lambda dA \tag{12}$$

The satellite irradiance of the satellite target at the entrance of the detector is

$$E = \sum_n E_n \tag{13}$$

In summary, the spectral irradiance of any point on the LEO satellite to the entrance pupil plane is calculated by the structural characteristics of the space object, the material properties and the parameters of the orbital characteristics. Structural properties refer to geometric models (cubes, cubes, cylinders, and circles). Material properties refer to surface reflectivity, surface root mean square roughness, and surface autocorrelation length. The orbital characteristics refer to the relative positional relationship between the radiation source, the target satellite and the observation satellite. Figure 4 shows a simple geometric model of imaging detection.

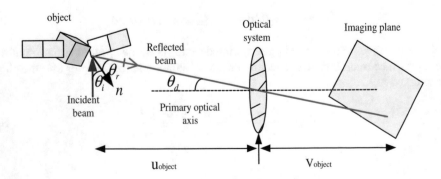

Fig. 4. Geometric model of imaging detection

3 Simulation of Visible Light Scattering Characteristics of Space Object

3.1 Target Parameters

(1) Structural characteristics
The body size of target is 2000 mm × 2000 mm × 2500 mm, and the size of solar panels is 2200 × 2700 mm.

(2) Material properties
The material of the target surface is aluminum. The surface mean square roughness σ is 0.2 μm, the surface autocorrelation length is 5.89 μm, and the reflectivity is 0.81. The material of the solar windsurfing is the cell sheet. The surface root mean square roughness σ is 0.05 μm, the surface autocorrelation length is 2 μm, and the reflectivity is 0.3.

(3) Orbital characteristics
The movement of the sun, the moon, the observation satellite and the target satellite relative to the earth follows the Kepler's law, and the commonly used elliptical equations of motion are expressed as:

$$M = (t - \tau)\sqrt{1/a^3} \tag{14}$$

$$
\begin{aligned}
v = M &+ e(2 - e^2/4 + 5e^4/96)\sin M + e^2(5/4 - 11e^2/24)\sin 2M \\
&+ e^3(13/12 - 43e^2/64)\sin 3M + 103e^4\sin 4M/96 + 1097e^5\sin 5M/960
\end{aligned}
\tag{15}
$$

$$r = \frac{a(1 - e^2)}{1 + e\cos v} \tag{16}$$

where a is the orbital half-long axis, e is the orbital eccentricity, t is the observation time, and τ is the past point of time. J2000.0 standard calendar from the beginning of the Julian day is used.

The calculation process of the target characteristics is shown in Fig. 5.

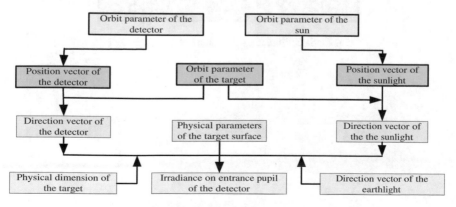

Fig. 5. Flow chart of target characteristics calculation

3.2   Simulation of LEO Satellite Image Based on OpenGL

The imaging simulation is implemented using OpenGL scene rendering. There are four parts: (1) Introduction of the target model file. (2) The lighting environment settings. (3) The camera parameter settings. (4) The final rendering simulation (Figs. 6 and 7).

3.3   The Actual Imaging Simulation Results

The actual process of imaging can be regarded as the degradation results of ideal imaging, including the following factors [10]: (1) Target movement. (2) Camera movement. (3) Image shift blur caused by the jitter. (4) Defocus blur caused by inaccurate control of camera focal length. (5) Fuzzy and internal dark Current noise caused by camera system performance.

The degradation process can be simulated by the corresponding simulation image. According to the above simulation principle and the principle of degradation to the image simulation, the simulation result is shown below:

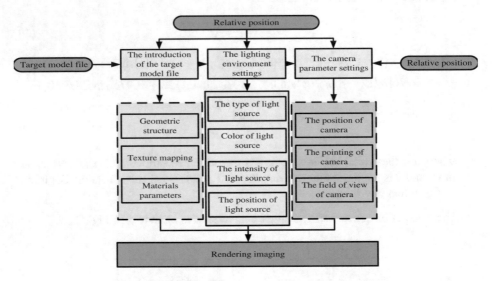

Fig. 6. The simulating flowchart of imaging model

Fig. 7. The actual imaging simulation results

4 Conclusion

In this paper, we introduce the Davies-BRDF model to describe the visible light scattering characteristics of the surface of the LEO satellite accurately. According to the given structural characteristics, material properties and orbital characteristics, the structure model of the LEO satellite is established. The target surface is decomposed in regions and meshed in grids based on the shape and material. Each surface element is analyzed and superimposed to establish the LEO satellite visible light scattering characteristic model of Davies-BRDF. 3D object for a given LEO satellite is simulated using OpenGL tool. The simulation process includes model file import, illumination environment setting and camera parameter setting. Finally the actual imaging simulation result is obtained by superimposing these degradation factors on the simulation result. Experimental results show that the Davies-BRDF model can simulate the spatial distribution of the reflected light of the space object more accurately, and it is suitable for a variety of surface materials. It also improves the realism and accuracy of the imaging effect of the space imaging camera, and provides theoretical foundation and technical support for the schematic design and performance analysis of space-based imaging detection systems.

References

1. Sharma, J.: Space-based visible space surveillance performance. J. Guid. Control Dyn. **23**, 154–159 (2000)
2. Hu, J.: Research on Monte Carlo Ray tracing based imaging effect simulation of space camera. University of Chinese Academy of Sciences (2013)
3. Guo, M., Wang, X.: IR modeling and simulation of LEO satellite/star and space environment. Infrared Laser Eng. **39**, 399–404 (2010)
4. Sun, C., Yuan, Y., Zhang, X.: Modeling of infrared characteristics of deep LEO satellite. Acta Phys. Sin. **59**, 7523–7530 (2010)

5. Claustrea, L., Boucher, Y., Paulin, M.: Wavelet-based modeling of spectral bidirectional reflectance distribution function data. Opt. Eng. **43**, 2327–2339 (2004)
6. Pont, S.C., Koenderink, J.J.: Bidirectional reflectance distribution function of specular surfaces with hemispherical pits. J. Opt. Soc. Am. A Opt. Image Sci. Vis. **19**, 2456–2466 (2002)
7. Donald Bédard, M., Lévesque, M., Wallace, B.: Measurement of the photometric and spectral BRDF of small Canadian satellites in a controlled environment. In: AMOS, pp. 1–10 (2011)
8. Prokhorov, V., Hanssen, L.: Algorithmic model of microfacet BRDF for Monte Carlo calculation of optical radiation transfer. Proc. SPIE **5192**, 141–157 (2003)
9. Yuan, Y., Sun, C., Zhang, X.: Measuring and modeling the spectral bidirectional reflection distribution function of LEO satellite surface material. Acta Phys. Sin. **59**, 2097–2103 (2009)
10. Glass, W., Duggin, M.J., Motes R.A.: Multi-spectral image analysis for improved space object characterization. Proc. SPIE 7467 (2009)

Analysis of Elevation Precision for Small Baseline Stereovision

Jinping He[✉], Yuchen Liu, Bin Hu, Yingbo Li, and Haibo Zhao

Key Laboratory for Advanced Optical Remote Sensing Technology of Beijing,
Beijing Institute of Space Mechanics and Electricity, Beijing 100094, China
13671364193@163.com

Abstract. Aiming at the optimal problem of selecting the ratio of baseline to satellite altitude in stereovision, a novel elevation precision model which is considering occlusion is proposed. The model is obtained through calculating the average elevation error of occlusion image areas and no-occlusion image areas. The elevation error in the no-occlusion areas can be represented through maximizing a local similarity coefficient. And the elevation error in the occlusion image areas can be estimated through averaging a certain point elevation. This model use is not only to select the optimal ratio, but also to analyze error sources from three dimensional characteristics of ground objects, the noise level of imaging system, and stereo-matching algorithms. The simulation experiment and the physical camera experiment all proof that there was an optimal B/H value in the small baseline stereovision. The optimal B/H values were respectively 0.1 and 0.03. The simulation experimental results represent that the higher the height of the highest building was, the bigger the elevation error was. This regularity also tallied with the precision model.

Keywords: Small baseline · Elevation precision · Stereovision
No-occlusion and occlusion areas · Error sources · Optimal ratio

1 Introduction

The camera optical center line of two observation points is known as the baseline in stereo imaging. The distance between the optical centers is the baseline length denoted by B. The ratio of the baseline length to the satellite altitude is represented as B/H. The measurement H is relative to the height datum plane. For a given optical camera, the B/H directly determines the elevation precision of stereovision under the condition of the same algorithm for image matching.

It is traditionally required that the baseline separation between images (or the base-to-height ratio) be very large in order to ensure the largest image disparity range for effective measurement. Currently, a B/H ratio in the range of 0.3–1.0 is preferred in the high resolution commercial satellite which has the capabilities of surveying and mapping. And the elevation estimation is mainly obtained from col-linearity equation. The equation can derive that the wide baseline setting of the stereo camera can improve the accuracy of the reconstruction elevation compared with the small baseline setting. Delvit J.M. utilized the SPOT5 images to conduct the small baseline experiment, and

© Springer Nature Switzerland AG 2018
H. P. Urbach and Q. Yu (Eds.): ISSOIA 2017, SPPHY 209, pp. 199–207, 2018.
https://doi.org/10.1007/978-3-319-96707-3_22

the elevation precision was better than one meter if the B/H ratio was in the range of 0.12–0.84 [1]. Julie D. from France's national space research center proposed the elevation error model of a certain point through maximizing a local similarity coefficient [2]. The elevation precision in function of B/H didn't have the first-order zero crossing. There was a contradiction between the above feature and the experimental result in which B/H has a minimal value around B/H = 0.1. Gareth presented the very narrow baseline stereo (e.g. less than 0.01) [3]. He considered that the narrow baselines could decrease the possibility of building occlusion.

A large B/H also means more changes between the images (more different hidden surfaces, differences in radiometry, larger geometrical deformations, moving objects, etc.), hence more difficulties in the matching process. The occlusion differences with a large B/H are much more critical than with a small coefficient. The method of small baseline is very suitable for the urban areas, where the depth can change very fast. The coupled B/H in occlusion areas is decomposed from the 3D information of surface features in this paper. The elevation precision model of occlusion areas in function of B/H is firstly constructed. The novel model is proposed through calculating the average elevation error of occlusion image areas and no-occlusion image areas. The simulation experiment and the physical camera experiment all proof the correctness of the elevation model.

2   The Principle of Stereo Imaging with Small Baseline

The principle of stereo imaging with small baseline is shown as Fig. 1. The focal length of an imaging camera in H orbit altitude is f and its pixel size is p. The camera snapshots at different positions with B baseline toward the ground points A and C. The line O_1A parallels to the auxiliary line O_2D. According to the similarity of ΔO_2AD and $\Delta O_2a_{11}a_2$, the difference of coordinate values of point A in different images can be calculated as

$$p_a = a_1 - a_2 = a_{11} - a_2 = \frac{fB}{H - Z_A},\tag{1}$$

where the image coordinate values a_1, a_2, c_1 and c_2 are the measurements related to the camera optical centers and the heights H, Z_A and Z_C are related to the reference plane.

In the same way, the difference of coordinate values of point C in different images can be obtained as

$$p_c = c_1 - c_2 = \frac{fB}{H - Z_C},\tag{2}$$

Based on the formulas (1) and (2), the relative height difference can be got by

$$\Delta Z = Z_A - Z_C = \frac{\Delta p(H - Z_C)}{p_a},\tag{3}$$

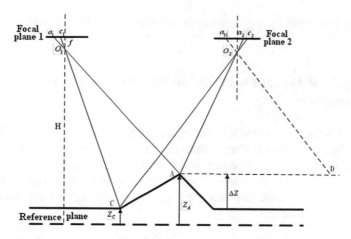

Fig. 1. The principle of stereo imaging

where $\Delta p = p_a - p_c = (a_1 - c_1) - (a_2 - c_2)$ represents the visual disparity. If the imaging with space remote sensor faces the general landform ($H \gg Z_C$), the formula (3) can be simplified as

$$\Delta Z \cong \Delta p \frac{H^2}{fB} = \Delta p \times \frac{H}{f} \times \frac{H}{B} = \frac{\Delta p}{B/H} \times \frac{H}{f}. \tag{4}$$

According to formula (4), the absolute elevation of two ground points is proportionate to the disparity of image pair.

Then the relative elevation in the image plane can be calculated by the formula

$$\Delta Z \cong \frac{\Delta p}{B/H} \tag{5}$$

From the standpoint of image analysis, the importance of formula (5) is the fact that the elevation difference between two ground points is proportionate to their disparity. Comparing with the traditional approach of col-linearity equation, the small baseline stereovision is not with complicated model and just needs to make the image pair matching accurately. This method can satisfy the engineering requirements with the sub-pixel precision of image matching. So the matching calculation is the key process in the small baseline stereovision.

The precision of traditional matching method for stereovision is in the integer pixel level. It can satisfy the precision requirement for the small baseline stereovision. So the sub-pixel matching algorithms are proposed [5, 6]. In this paper, the local phase correlation algorithm is adopted [7]. This approach is based on the shift theory of Fourier transformation. The relative displacement of the similar image pair can be represented by the phase difference of their Fourier transformation. The resampling along the epipolar lines can make the image pair have one direction disparity. Then using the disparity value along one direction, the relative elevation can be computed by the formula (5).

3   Analysis of Elevation Precision for Small Baseline Stereovision

The camera carrying on the satellite image toward the same ground area from different positions. Then the gathering images form the image pair. The overlap region of image pair can be divided into occlusion areas and no-occlusion areas.

The elevation error of the caught overlap region in the image plane can be obtained through averaging the elevation error between occlusion image areas and no-occlusion image areas. It follows that

$$e = \frac{1}{M \times N} \cdot \left(\sum_i \sum_{x \in \Omega_{occlu}} e^i_{occlu}(x) + \sum_{x \in \Omega_{nocclu}} e_{nocclu}(x) \right), \qquad (6)$$

where $M \times N$ is the image size of the overlap region, Ω_{occlu} represent the occlusion image areas, and Ω_{nocclu} represent the no-occlusion image areas.

The elevation error $e_{nocclu}(x_0)$ of a certain point x_0 in the no-occlusion areas through maximizing a local similarity coefficient [2] can be represented as

$$e_{nocclu}(x_0) \leq \left| Z_{real}(x_0) - \frac{\left\langle d^{\tau_m u}_{x_0}, Z_{real}(x_0) \right\rangle_{\varphi_{x_0}}}{\left\langle d^{\tau_m u}_{x_0}, 1 \right\rangle_{\varphi_{x_0}}} \right| \\ + \frac{\sigma_b \|g\|_{L^2}}{B/H \cdot \|\tau_m u'\|_{\varphi_{x_0}} \left(\left\langle d^{\tau_m u}_{x_0}, 1 \right\rangle_{\varphi_{x_0}} \right)^{1/2}}, \qquad (7)$$

where $Z_{real}(x_0)$ denotes the real elevation in the image plane at x_0, $d^{\tau_m u}_{x_0} = \frac{\|\tau_m u\|^2_{\varphi_{x_0}} \tau_m u'^2 - \langle \tau_m u, \tau_m u' \rangle_{\varphi_{x_0}} \tau_m u \tau_m u'}{\|\tau_m u\|^4_{\varphi_{x_0}}}$, $\tau_m u(x) = u(x+m)$, $u(x)$ is one image of the stereoscopic pair at point x, m is the measurement disparity, $\langle h_1 - h_2 \rangle_{\varphi_{x_0}} = \int \varphi_{x_0}(x) h_1(x) h_{2(x)} dx$, $\|h\|_{\varphi_{x_0}} = \sqrt{\int \varphi_{x_0}(x) h^2(x) dx}$, $\varphi_{x_0}(x)$ is the shifted function of the correlation window at x_0, u' denotes the one-sided directional derivative of u, σ_b is the standard deviation of the Gaussian noise b, and $\|g\|_{L^2}$ is the weighted norm for the spread function g.

One certain point in the occlusion image areas can be estimated through averaging. The elevation error model of a certain point in occlusion can be represented as

$$e^i_{occlu}(x) = \frac{1}{2} Z^i_{real} \qquad (8)$$

where i denotes the i th building, x is a certain point in occlusion, Z^i_{real} denotes the real elevation of the i th building in the image plane.

The area ratio between the occlusion and the no-occlusion is required in order to calculate the average error of the two areas. So the size of occlusion areas is firstly derived. The length of occlusion areas at X direction can be obtained from the similar triangles $\Delta OO1O2$ and ΔOAC in Fig. 1. Therefore, there is X1 + X2 $\approx$ X3 + X4 = B/H·Z_{real} where Z_{real} is the real elevation of the higher building in Fig. 2.

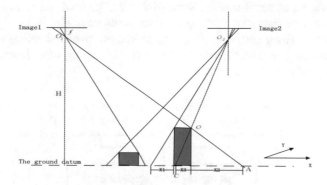

Fig. 2. A diagram of occlusion phenomenon

Then the size of occlusion areas of the i th building is

$$s^i \cong \frac{B}{H} \cdot Z_{real}^i \cdot x^i + \frac{B}{H} \cdot Z_{real}^i \cdot y^i, \tag{9}$$

where x^i, y^i are separately represented building lengths of the X, Y directions of Cartesian coordinates in the occlusion areas of image plane.

The formulas (7), (8) and (9) are substituted into the formula (6), then

$$\begin{aligned}
e &= \sum_i \frac{S^i}{M \times N} \cdot \frac{1}{2} \Delta Z^i + \frac{1}{M \times N} \sum_{x \in \Omega_{nocclu}} e_{nocclu}(x) \\
&\approx \frac{B}{H} \cdot \sum_i \frac{\left(Z_{real}^i\right)^2 (x^i + y^i)}{M \times N} \\
&\quad + \frac{1}{M \times N} \sum_{x \in \Omega_{nocclu}} \left| Z_{real}(x) - \frac{\langle d_x^{\tau_m u}, Z_{real}(x) \rangle_{\phi_x}}{\langle d_x^{\tau_m u}, 1 \rangle_{\phi_x}} \right| \\
&\quad + \frac{1}{B/H} \cdot \sum_{x \in \Omega_{nocclu}} \frac{\sigma_b \cdot \|g\|_{L^2}}{M \times N \cdot \|\tau_m u'\|_{\phi_x} \left(\langle d_x^{\tau_m u}, 1 \rangle_{\phi_x} \right)^{1/2}},
\end{aligned} \tag{10}$$

Assuming that the B/H ratio is the independent variable, the final elevation error in function of B/H is not a monotonic function from the formula (10). There is a first-order zero crossing in this error model. There is an optimal B/H value if minimizing the final elevation error. If the elevation error is relatively large values, the characteristics of surface features plays a leading role in the elevation error. And if the elevation error is relatively small, the noises play a leading role in the elevation error.

4 Experimental Results

Simulation process of small-baseline stereovision can be obtained from reference [8]. The calculation flow chart of the precision errors is shown in Fig. 3. Firstly two images will be resampled along epipolar lines. Then match the two resampling images in the frequency domain. The disparity map is obtained. Comparing with the real elevations, the precision errors can be calculated.

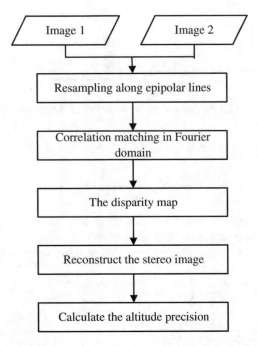

Fig. 3. The calculation flow chart of the precision errors

Under the condition of a certain degree of noise (the noise standard deviation is equal to 0.003), the elevation errors of the same scene for different B/H values are firstly calculated. As shown in Fig. 4, the highest building in Scene 1 is 45 m, and there are three 18.36 m buildings and one 37.5 m building. Figure 5 shows the imaging pair of Scene 1 when B/H = 0.5. The minimum elevation error is 3.336 pixels and the corresponding optimal B/H is equal 0.1. This phenomenon tallies with the formula (10). With regard to the same scene, the B/H ratio has an optimal value. And the B/H values are not as big as possible or as small as possible (Table 1).

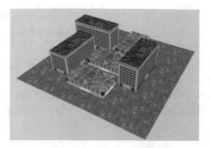

Fig. 4. Stereoscopic Scene 1

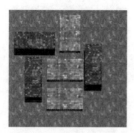

Fig. 5. The stereoscopic pair of Scene 1 when B/H = 0.5

Table 1. The list of elevation errors of Scene 1 corresponding to the different B/H

B/H	0.01	0.02	0.03	0.05	0.1	0.2	0.3	0.5
Elevation error (pixels)	10.736	8.378	6.613	6.174	3.336	5.325	6.369	14.955

Under the condition of the same noise, other two scenes are synthetically analyzed when B/H = 0.1. There are two buildings in Scene 2 and one building in Scene 3. The highest building in Scene 2 is 30 m, and in Scene 3 is 9.12 m. As shown in Table 2, the experimental results represent that the higher the height of the highest building is, the bigger the elevation error is. This regularity tallies with the formula (10). The elevation error is proportional to the elevation square of surface features from first term of the error expression. In other words, the highest terrestrial object influences the elevation error most.

Table 2. The list of elevation errors corresponding to the different highest buildings

Scenes' name	Scene 1	Scene 2	Scene 3
The height of highest buildings (m)	45	30	9.12
Elevation error (pixels)	3.336	1.629	0.877

Finally, the elevation error model is further verified through the physical test with the stereo target as shown in Fig. 6. The real heights of the objects on the target are measured by laser altimeter. The imaging distance is 30 m. F number of the camera is 16. The pixel size of the detector is 9 μm. The Fig. 7 is the disparity map of stereo target obtained by correlation methods in Fourier domain [5]. The reconstructing result of the stereo target is shown in Fig. 8. When B/H = 0.03, the minimal elevation error is 5.71 pixels. The result accords with the formula (10) (Table 3).

 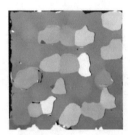

Fig. 6. Stereo target   **Fig. 7.** The disparity map of
                                  stereo target

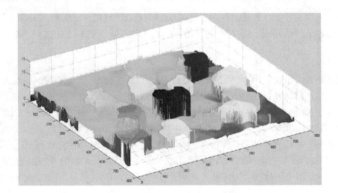

Fig. 8. The stereo reconstructing result for the target

Table 3. Elevation errors of the stereo target corresponding to the different B/H

B/H value	0.01	0.03	0.05	0.08	0.1	0.12
Elevation error (pixels)	8.6244	5.7132	6.1212	7.4304	8.586	8.7492

5   Conclusion

The previous elevation model varies monotonically with the B/H ratio. This is not in accordance with the experimental results and the current practical applications. The elevation precision in function of B/H in this letter which considers occlusions has the first-order zero crossing. And it is in coincidence with the simulation and experimental results. This model can be used to select the B/H ratio and analyze the elevation error source. Further obtained from the proposed elevation model, the small baseline stereovision is more suitable for city areas in which the building objects are with a high density. The method with small baseline can compute the elevations of the whole overlap region at once. And it can improve the efficiency of the data acquisition. An economic and feasible means also was provided for the large scale mapping of space stereo.

References

1. Delvit, J.M.: Small baseline, data acquisitions and applications. In: ICSO Congress Toulouse (2008)
2. Julie, D., Rougé, B.: Small baseline stereovision. J. Math. Imaging Vis. **28**, 209–223 (2007)
3. Gareth, L.K.M., Liu, J.G., Yan, H.: Precise subpixel disparity measurement from very narrow baseline stereo. IEEE Trans. Geosci. Remote Sens. **48**(9), 3424–3433 (2010)
4. Robert, A.S.: Remote Sensing: Models and Methods for Image Processing, 3rd edn. Publishing House of Electronics Industry, Beijing (2010)
5. Morgan, G.L.K., Liu, J., Yan, H.: Sub-pixel stereo matching for DEM generational from narrow baseline stereo imagery. In: IEEE Proceedings of International Geoscience and Remote Sensing Symposium, Boston, pp. 1284–1287 (2008)
6. Morgan, G.L.K., Liu, J., Yan, H.: Precise subpixel disparity measurement from very narrow baseline stereo. IEEE Trans. Geosci. Remote Sens. **48**, 3424–3433 (2010)
7. Liu, Y., He, J., Hu, B., et al.: Simulation analysis of small-baseline stereo surveying. Spacecr. Recov. Remote Sens. **37**, 95–101 (2016)
8. Sébastien, L., Sylvain, B., François, A., et al.: Automatic and precise orthorectification, coregistration, and subpixel correlation of satellite images, application to ground deformation measurements. IEEE Trans. Geosci. Remote Sens. **45**, 1529–1558 (2007)

Technique of Polarization Remote Sensing on GEO and Quick Check Method of Polarization Image

Zhuoyi Chen$^{(\boxtimes)}$, Fengjing Liu, Xianghao Kong, and Guo Li

Beijing Institute of Spacecraft System Engineering, Beijing, China
chenzhuoyi541l@sina.com

Abstract. Polarization signal is an important part of the reflected signal from the earth's surface. As different object and places reflect different state of polarization, the feature information of objects can be easier distinguished by using polarization remote sensing comparing the traditional remote sensing technique. On the other hand, staring imaging mode of polarization sensing on geostationary orbit cannot be limited by arcs position as it is used on low orbit, which can provide a better sensing effect; therefore multiple advantages can be taken by using polarization remote sensing on GEO. And the efficiency of polarization sensing could be significant improved, when using quick check method of polarization image to analyze a best degree of polarization at diverse solar altitude angle. In this article, a principle and technology of the quick check method of different polarization image is analyzed and demonstrated under the condition of space observation at different solar altitude angle, and a simulation method is used to validate the effectiveness of quick check method of polarization image. A technical support of space polarization remote sensing, especially the sensing system on GEO, will be provided.

Keywords: GEO · Polarization remote sensing
Quick check method of polarization image

1 Introduction

Polarization is a fundamental property of light. Different structures or different states of the same ground produce different polarization state. During the process of interacting with the object, as the structure of the surface, the internal structure and the angle of incidence are different, not only its intensity changes, but also its polarization state tend to change [1]. Polarization measurements provide better feature information that is easy to distinguish between targets than non-polarized measurements. In the geostationary orbit application of polarized remote sensing technology, due to low rail observation arc limit, the same scene of the more selective degree of polarization, better basis for comparative analysis, and thus can get better polarization imaging quality. Due to the low number of polarized detectors currently in orbit, the spatial resolution is low, and can only serve the detection of cloud and aerosol parameters, it is difficult to support the identification of objects. Therefore, it is necessary to analyze the best degree of polarization by using the multi-polarization image detection method by means of

© Springer Nature Switzerland AG 2018
H. P. Urbach and Q. Yu (Eds.): ISSOIA 2017, SPPHY 209, pp. 208–214, 2018.
https://doi.org/10.1007/978-3-319-96707-3_23

ground simulation analysis, and to maximize the geostationary effect of the geostationary orbit and improve the polarized observation efficiency of the satellite.

2 Fundamentals of Polarization Remote Sensing

The sunlight is incident on the atmosphere, resulting in scattering, which has a certain degree of polarization. Sunlight or through the atmosphere of the sky light to the ground objects, reflected by the objects, refraction or scattering of electromagnetic radiation will also become a certain polarization of light waves, their characteristics are mainly expressed in: perpendicular to the light on the plane, light in all directions the energy distribution is not uniform, there will be polarization phenomenon. Surface or atmospheric target objects, in the process of reflection, scattering and transmission of electromagnetic radiation, will produce by their own nature of the polarization characteristics [2]. Polarized light is providing new means for remote sensing target recognition. Experiments show that the polarization information in the soil vegetation classification, aerosol detection and seawater surface state research has it's superiority.

3 Characteristics of Static Orbit Polarization Remote Sensing

The application of polarized remote sensing technology in geostationary orbit has a prominent advantage over the low-range remote sensing field, which is the static gaze imaging mode is not limited by the low-orbit observation arc and can provide more choices for the observed polarization and thus the best degree of polarization can be analyzed by using a set of multi-polarization images at different solar elevation angles, which can improve the polarized observation performance of the satellite to the maximum extent.

4 The Influence of Observation Angle on Polarization Detection

In the simulation analysis of this project, the solar zenith angle 54.36° and the azimuth angle 120.43° were calculated under the same atmospheric conditions and sea surface model, and the relationship of atmospheric reflectivity, atmospheric polarization reflectance and atmospheric top were calculated respectively. It is concluded that the apparent reflectance and the apparent polarized reflectance of the atmospheric top, both the atmospheric reflectance, the atmospheric polarization reflectance, the Amoy table model and the atmospheric coupling, all change with the observation direction. In the vicinity of the specular reflection direction, the apparent reflectance and apparent polarized reflectance increase of the atmospheric reflectivity and the atmospheric reflectance, atmospheric reflectance and atmospheric polarization reflectance are more significant, and the maximum apparent reflectance of the atmospheric top is close to 10 times. In the non-specular reflection direction, atmospheric reflectivity, atmospheric polarization reflectance and atmospheric apparent reflectivity, the apparent polarization reflectivity is not very different. This indicates that the observation direction has a great influence on the polarization detection. To the marine polarization detection, special attention should be paid to the influence of specular reflection and its application.

5 Rapid Polarization Detection Technology

5.1 Analysis of the Principle of Polarization Detection

The target background polarization characteristic simulation is actually simulated unbiased solar direct light through the atmospheric scattering absorption, the target reflection and the background environment after the process of reflection into polarized light. The polarized two-way reflection characteristic of the target is usually expressed in terms of the p-BRDF [3], which can be expressed by a 4 * 4 matrix $\mathbf{F_r}(\theta_i, \phi_i; \theta_r, \phi_r; \lambda)$.

$$F_r = \begin{pmatrix} f_{00} & f_{10} & f_{20} & f_{30} \\ f_{01} & f_{11} & f_{21} & f_{31} \\ f_{02} & f_{12} & f_{22} & f_{32} \\ f_{03} & f_{13} & f_{23} & f_{33} \end{pmatrix} \tag{5.0}$$

The transmission relationship of the incident radiation to the reflected radiation is expressed by the Muller matrix M, and the reflected Stokes vector I_r is obtained from the incident Stokes vector I_i of the material interface.

$$\begin{pmatrix} I^r \\ Q^r \\ U^r \\ V^r \end{pmatrix} = \begin{pmatrix} M_{00} & M_{10} & M_{20} & M_{30} \\ M_{01} & M_{11} & M_{21} & M_{31} \\ M_{02} & M_{12} & M_{22} & M_{32} \\ M_{03} & M_{13} & M_{23} & M_{33} \end{pmatrix} \begin{pmatrix} I^i \\ Q^i \\ U^i \\ V^i \end{pmatrix} \tag{5.1}$$

The relationship between the Mull matrix M and the matrix Fr in the above equation is as follows:

$$\mathbf{M} = \begin{pmatrix} \int f_{00} \cos \theta_r d\Omega_r & \int f_{01} \cos \theta_r d\Omega_r & \int f_{02} \cos \theta_r d\Omega_r & \int f_{03} \cos \theta_r d\Omega_r \\ \int f_{10} \cos \theta_r d\Omega_r & \int f_{11} \cos \theta_r d\Omega_r & \int f_{12} \cos \theta_r d\Omega_r & \int f_{13} \cos \theta_r d\Omega_r \\ \int f_{20} \cos \theta_r d\Omega_r & \int f_{21} \cos \theta_r d\Omega_r & \int f_{22} \cos \theta_r d\Omega_r & \int f_{23} \cos \theta_r d\Omega_r \\ \int f_{30} \cos \theta_r d\Omega_r & \int f_{31} \cos \theta_r d\Omega_r & \int f_{32} \cos \theta_r d\Omega_r & \int f_{33} \cos \theta_r d\Omega_r \end{pmatrix} \tag{5.2}$$

The influence of the atmospheric environment is calculated by the classical vector atmospheric radiation model [4]. According to the target background polarization characteristic model and the atmospheric polarization characteristic transmission model, after considering the target background and the atmospheric coupling effect, the target reflected light background environment reflected light, atmospheric scattered light and multiple scattered light between target background and the atmosphere to reach the polarization detector along the simulation of the process to reach the probe into the pupil of the atmospheric information can be expressed by the vector equation:

$$\mathbf{I}_\lambda^{app}(\theta_s, \theta_v; \varphi) = \mathbf{I}_\lambda^{Atm}(\theta_s, \theta_v; \varphi) + \frac{\mathbf{I}_\lambda^{suf}(\theta_s, \theta_v; \varphi)}{1 - \mathbf{I}_\lambda^{suf}(\theta_s, \theta_v; \varphi) \cdot S} T_\lambda(\theta_s) T_\lambda(\theta_v) \qquad (5.3)$$

The relationship between the I matrix and the Stokes parameter is as follows:

$$\mathbf{I} = \begin{pmatrix} \pi I / \mu_0 F_0 \\ \pi Q / \mu_0 F_0 \\ \pi U / \mu_0 F_0 \\ \pi V / \mu_0 F_0 \end{pmatrix} \qquad (5.4)$$

In the above formula, λ is the wavelength, θ_s is the solar zenith angle, θ_v is the observation zenith angle, φ is the relative azimuth angle; $\mathbf{I}_\lambda^{app}(\theta_s, \theta_v; \varphi)$ is the normalized Stokes vector at the entrance of the sensor; $\mathbf{I}_\lambda^{Atm}(\theta_s, \theta_v; \varphi)$ Is the normalized Stokes vector generated by the scattering of atmospheric molecules and aerosols; $\mathbf{I}_\lambda^{suf}(\theta_s, \theta_v; \varphi)$ is the normalized Stokes vector produced by the target background; $T_\lambda(\theta_s)$ is the incident direction (the sun-target path) The total scattering transmittance, $T_\lambda(\theta_v)$ is the total scattering transmittance in the observation direction (target-observation point path); S is the atmospheric hemisphere reflectivity, μ_0 Is the cosine of the sun's zenith angle, and the F_0 layer is the solar irradiance outside the atmosphere.

$$T_\lambda(\theta_s) = e^{-\tau/\mu_s} + t_d(\mu_s) \qquad (5.5)$$

$$T_\lambda(\theta_v) = e^{-\tau/\mu_v} + t_d(\mu_v) \qquad (5.6)$$

In the above formula, μ_s is the cosine of the sun's zenith angle, μ_v is the cosine of the observation zenith angle; $e^{-\tau/\mu_s}$, $e^{-\tau/\mu_v}$ are the direct transmittance in the direction of the incident direction and the observation direction respectively; $t_d(\mu_s)$ and $t_d(\mu_v)$ are the scattering transmittance in the direction of the incident direction and the observation direction respectively. The scattering transmittance mainly comes from the scattering of atmospheric molecules and aerosols, including multiple scattering of the surface and the atmosphere. The size of the scattering transmittance depends mainly on the aerosol model because the formula for the calculation of atmospheric scattering is determined.

In summary, the normalized Stokes vector $\mathbf{I}_\lambda^{suf}(\theta_s, \theta_v; \varphi)$ generated by the target background in Eq. (5.1) can be measured in real terms; the other three terms, such as $T(\theta_s)$, $T(\theta_v)$ and $\mathbf{I}_\lambda^{Atm}(\theta_s, \theta_v; \varphi)$, are related to atmospheric scattering. Can be accurately calculated using the atmospheric vector radiation transmission equation.

The reflectivity ρ and the polarization reflectivity ρ_P at the entrance of the sensor are related to the normalized Stokes parameter I, Q, U, V at the entrance of the sensor:

$$\begin{cases} \rho = \frac{\pi I}{\mu_0 F_0} \\ \rho_P = \frac{\pi \sqrt{Q^2 + U^2 + V^2}}{\mu_0 F} \end{cases} \qquad (5.7)$$

According to the range and spectral response function of each band of the detector, the polarized reflectivity $\mathbf{I}_{p,\lambda}^{app}(\theta_s, \theta_v; \varphi)$ at the entrance of the sensor can be obtained from the formula (5.3) and the formula (5.7), which can realize the optical polarization considering the target background and atmospheric coupling Imaging link.

5.2   Polarization Image Conversion

If the polarimetric reflectance of the same kind of object are the same, the ground polarized reflectance image can be generated by mapping the measured polarized reflectance of the feature to the corresponding region in the figure [5]. However, due to the non-uniformity of the target object itself, the same type of objects in different regions of the polarization reflectivity is not exactly the same. In order to reflect this nonuniformity, it is further assumed that the change in the reflectance of the same type of object is proportional to the change in the reflectivity itself. Whereby each pixel in the ground polarized reflectance image can be calculated using the formula (5.4).

$$\rho_{P,\lambda}^{Suf}(i,j; \theta_s, \theta_v, \varphi) = \frac{\rho_{P,\lambda}^{Meas,k}(\theta_s, \theta_v, \varphi) \cdot DN^k(i,j)}{DN_{avg}^k} \tag{5.8}$$

In the above formula, λ denotes the wavelength, k is the feature type, θ_s is the solar zenith angle, θ_v is the observation zenith angle, and φ is the relative azimuth angle. $\rho_{P,\lambda}^{Suf}(i,j; \theta_s, \theta_v, \varphi)$ is the polarization reflectance of the pixel (i, j) in the ground polarized reflectance simulation image; $\rho_{P,\lambda}^{Meas,k}(\theta_s, \theta_v, \varphi)$ is the polarized reflectance value of the feature k measured in a certain observation direction; $DN^k(i,j)$ Is the gray value of the pixel (i, j) in the intensity image, and DN_{avg}^k is the gray scale of the feature k in the intensity image.

5.3   Polarization Image Simulation

According to the above principle, $T(\theta_s)$, $T(\theta_v)$ and $\mathbf{I}_\lambda^{Atm}(\theta_s, \theta_v; \varphi)$ are calculated by 6SV, and the polarized reflectance image of the sensor is realized by the formula (5.1) and the formula (5.5). When calculating the amount of atmospheric radiation using 6SV, it can be obtained by setting the solar zenith angle, observing azimuth angle, atmospheric mode, aerosol model, observation height and band bandwidth.

The following Fig. 1(a) to (b) are the imaging simulation images when the zenith angle is 0°, 15°, 30°, 45° and 60°, respectively. It can be seen from the figure that when the zenith angle is 15°, the water body and the hull are the most difficult to distinguish. When the zenith angle is 60°, the contrast between the water body and the hull is the biggest.

According to the simulation method, the simulation results of polarization imaging under different weather conditions and observation azimuth angle can be analyzed. Through a series of objective image quality evaluation indexes, the polarimetric characteristics of the polar conditions under different conditions are compared, which provide the polarized detection and identification performance of the satellite in accordance with.

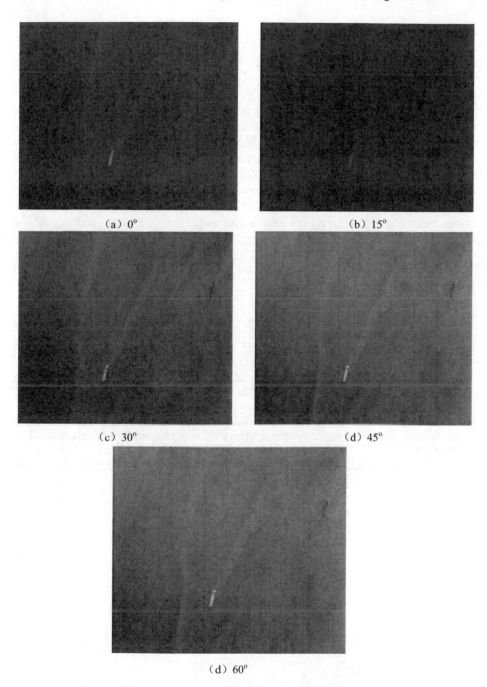

(a) 0°   (b) 15°

(c) 30°   (d) 45°

(d) 60°

Fig. 1. The simulation images when different zenith angle

6 Concluding Remarks

Polarization imaging detection technology is much more complicated than radiation imaging detection technology. The study of polarization imaging technology of space-based remote sensing system is only a single application in the initial stage, and the whole-link study of polarization imaging is also blank. This paper only explains the more preliminary work, suggesting that follow-up study can focus on the special treatment of polarized images. Ordinary illumination image processing methods have been very rich, and polarized images have their particularity, such as the general object of the degree of polarization is relatively small, is not conducive to human eye observation, and is not conducive to computer processing, so the polarized image has its special treatment. At the same time, the study of polarization image and ordinary image fusion algorithm can greatly improve the target recognition rate and expand the application of polarization imaging system.

References

1. Cairns, B., Travis, L.D., Russell, E.E.: The research scanning polarimeter: calibration and groundbased measurements. Proc. SPIE **3754**, 186–197 (1999)
2. Kopp, G., Lawrence, G., Rottman, G.: The total irradiance monitor design and on-orbit functionality. Proc. SPIE **5171**, 5171–5174 (2003)
3. Chimata, A.-P., Allen, C.: Development of an adaptive polarization-mode dispersion compensation system. ITTC-FY2003-TR-18834-02, Technical Report. The University of Kansas (2003)
4. Corbel, E., Lanne, S., Thiéry, J.P.: Improvement of first-order optical PMD compensator by relevant control of input state of polarization. In: OFC 2002, Anaheim, California, USA. OSA (2002). WQ3
5. Mishchenko, M.I., Geogdzhayev, I.V.: Satellite remote sensing reveals regional tropospheric aerosol trends. Opt. Express **15**, 7423–7438 (2007)

Research and Verification of Spatial Hyperspectral Data Processing Method

Yu Wang[✉] and Wei Nian

Key Laboratory for Advanced Optical Remote Sensing Technology of Beijing,
Beijing Institute of Space Mechanics and Electricity, Beijing, China
Cherishfriend@vip.sina.com

Abstract. In Chinese new generation meteorological satellite, Hyperspectral greenhouse gas monitoring instrument is the first spectral observation load dedicated to the spectral detection of greenhouse gases, with the ability to track the total amount of greenhouse gases such as CO_2, CH_4 and CO in orbit. In order to realize high precision and quantitative remote sensing monitoring of greenhouse gases, the monitor should have the characteristics of high signal to noise ratio (SNR) when the signal is weak. Therefore, increasing the effective spectral information and eliminating the noise to increase the SNR is an important research content of the system.

In this paper, a set of on-board data processing method is proposed, which can effectively control the noise of electronic circuits and improve the SNR. The validity of the method is verified by analyzing and processing the data of the monitoring instrument, which provides a new idea and direction for the subsequent satellite application of hyperspectral observation load.

Keywords: Hyperspectral · Data processing · Wavelet transform

1 Introduction

Hyperspectral greenhouse gas monitoring instrument (HGGMI) is the first spectral observation load of China dedicated to the spectral detection of greenhouse gases, with the ability to track the total amount of greenhouse gases such as CO_2, CH_4 and CO in orbit. The schematic diagram is as follows (Fig. 1):

When it is in orbit, the measured lights enter into the interferometer module through the two-dimensional pointing mirror. The input beam is interfered by the interferometer module, then get to the detector through the optical system compression and color convergence optical system. The modulated optical signals are converted to interferogram signals by the detector. The interferogram signal is amplified and filtered by the analog signal processing circuit and then fed into the ADC for digital quantization. The quantized signals are sent into the signal processing system to encode and then transmitted to the satellite data transmission system.

The interferogram signals are transmitted to the ground data center via satellite data. The ground software processing platform gives the spectral data to customers which based on the state of package and the original interferogram data, through

© Springer Nature Switzerland AG 2018
H. P. Urbach and Q. Yu (Eds.): ISSOIA 2017, SPPHY 209, pp. 215–226, 2018.
https://doi.org/10.1007/978-3-319-96707-3_24

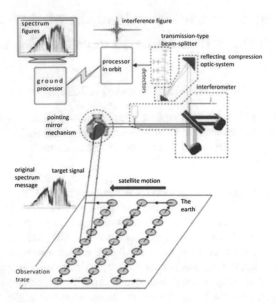

Fig. 1. The overall plan of Hyperspectral greenhouse gas monitoring instrument

Fourier transform spectrum restoration, through spectral calibration and radiometric calibration, and additional geographic information.

The channel model is shown in Fig. 2. The target radiation source obtains the corresponding digital signal by sampling through an optical channel, an electrical channel of a spectrometer. Among them, background noise is introduced in the optical channel, which is mainly caused by infrared radiation of the internal components of the spectrometer. In the electrical channel, a series of noise will be introduced related to detectors, such as photon noise, thermal noise, dark current noise, and quantization noise caused by sampling process.

The signal intensity of the space hyperspectral greenhouse gas monitor is very weak, and the spectrum is easily affected by the above noise factors, which leads to the decrease of the system signal-to-noise ratio. Therefore, before the establishment of calibration model, we should pretreat the spectral data, enhance the effective information of the spectrum, eliminate all kinds of uncertain factors, so that we can increase the SNR, improve the accuracy of spectral analysis. The results show that the SNR can be increased and the spectral analysis accuracy can be improved by wavelet denoising.

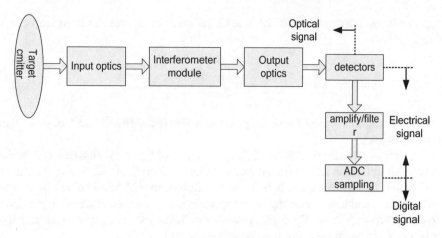

Fig. 2. Channel models

2   Principles and Methods of Wavelet Transform

Wavelet analysis is a new technique developed in recent 20 years, it is a new time domain analysis method developed based on functional analysis, harmonic analysis, numerical analysis, approximation theory and FFT analysis. Compared with Fourier analysis, wavelet analysis is a kind of analysis method with good localization in time domain and frequency domain.

Therefore, wavelet analysis can adaptively adjust the time-frequency window according to the needs of actual analysis, and can focus on any details of the signal in the time domain and frequency domain. It is called the microscope of signal analysis. At present, the theory and application of wavelet analysis has got rapid development, which has been widely applied in the field of numerical analysis, signal analysis, image processing, quantum theory, computer vision, pattern recognition, fault diagnosis and so on, it is considered a major breakthrough in methods and tools in recent years.

The locations of absorption peak, shape and peak intensity for the spectral (feature information localization) show the main features of greenhouse gas monitoring. Wavelet analysis is consistent with the local information extraction requirements, so the wavelet transform has become a powerful tool for spectral noise reduction.

2.1   Brief Introduction of Wavelet Transform Filter

Fourier analysis is the classical method for signal processing, but in Fourier analysis, the signal is completely spread out in frequency domain, and does not include any time-domain information. In contrast, wavelet analysis is a method of analysis with good localization in both time and frequency domains.

Suppose the spectrum measured is s(λ), its continuous wavelet transform can be defined as:

$$W_\Psi s(a,b) = \frac{1}{\sqrt{a}} \int\limits_{-\infty}^{+\infty} s(\lambda)\Psi(\frac{\lambda - b}{a})d\lambda \qquad (1)$$

In the formula, a stands for scaling factor, b stands for translation factor, λ stands for wavelength.

The wavelet transform function 公式 is the result of the translation of b and scaling of a on the wavelength axis of the mother wavelet function 公式. Choosing the wavelet function with specific local characteristics is optional, and the best choice can be made in the actual calculation according to the specific conditions. In practical applications, software processing is mostly digital processing. Thus, the discrete wavelet transform can be obtained by discretizing the continuous wavelet (1).

2.2 Research on Wavelet Noise Reduction Methods

There are many kinds of wavelet noise reduction methods, such as maximum detection, masking noise reduction and threshold de-noising. The thresholding method is the most commonly used as a noise reduction method, because its principle is relatively simple, and its calculation is small, and can effectively remove the noise and has been widely used in signal singularity at the same time. We use the method in this paper.

The main steps of threshold noise reduction can be divided into the following 3 steps:

a. Wavelet decomposition of the signal which is used to obtain the wavelet coefficients of the signal decomposition;
b. The wavelet coefficients are operated by thresholding, and the new wavelet coefficients are obtained;
c. The denoised signals are reconstructed from the new wavelet coefficients.

The process of wavelet threshold de-noising is discussed in detail.

First of all, the wavelet transform can be used to decompose the spectral signal with complex resolution, and the wavelet coefficients vector matrix is obtained as an example. Starting from the spectral signals at the top, the first order wavelet coefficients are decomposed to obtain the first order detail signal coefficients, and the first order approximate coefficients are obtained. Next, in expanding the scale of wavelet function, decomposition of the signal to approximate the first-order resolution scale, class repeat this process until the required date, finally we can get the bottom of the wavelet coefficient vector matrix. In the decomposition process, from the first step down, the signal decomposition resolution is getting lower and lower, and the corresponding frequency of the detail signal coefficients is gradually decreased until the lowest approximation of the first order of the approximate signal coefficients. It is assumed that the d1 wavelet coefficients that represent the highest coefficients are mainly corresponding to the noise in the original signal, while the a3 coefficients representing the lowest frequencies correspond to the smooth background spectrum in the original signal.

Then remove noise through thresholding for the wavelet coefficient. Generally speaking, spectral noise distributes in different scales, and the energy distribution is more concentrated in the small scale level, while the wavelet coefficients of spectral peaks corresponding will focus on a few middle scale. Therefore, noise can be removed effectively when we set the wavelet coefficients below the set threshold zero.

2.3   Introduction to Wavelet Function

Compared with FFT, wavelet function used in wavelet transform is not only, it can be said that all functions satisfy the admissible condition of wavelet can use as wavelet function. In the actual application, the wavelet basis function selection is a key problem, because using the different wavelet basis function analysis the same the problem will produce different results. In practical application, the wavelet basis function is selected according to the principle of self similarity. If the wavelet basis function is similar to the signal, the energy of the signal is more concentrated after wavelet transform. The wavelet basis function used in engineering mainly include HRRR, Daubechies, Marr, Symlet, Biorthorgonal, Morlet, Meyer wavelet and many others.

Different wavelet functions have different noise reduction effects. In this paper, three kinds of wavelet functions are used to analyze the effect of wavelet transform including Daubechies, Symlet, and Biorthorgonal.

3   Test Data Analysis

In third band, we take 50 sets of data using Labview to analyze the system signal-to-noise ratio before and after filtering. The specific mathematical expression of the signal-to-noise ratio is:

$$SNR_b(\sigma) = \frac{\langle \mathrm{Mean}_i\left(\left|S_{i,b}(\sigma)\right|\right)\rangle_b}{\langle \mathrm{Sdev}_i\left(\left|S_{i,b}(\sigma)\right|\right)\rangle_b} \tag{2}$$

In the expression, S is the obtained spectrum, b is the number of spectral section, i is the test spectrum number, Mean represents the average value of the multiple spectra, and Sdev represents the standard deviation of the test spectrum.

3.1   The Filtering Effect of Different Wavelet Basis Functions Adopted

3.1.1   Daubechies Wavelet

Daubechies wavelet is proposed by a famous wavelet analysis scholar Daubechies who studied the wavelet transform under the condition of 2 integer power. Daubechies series wavelets can be abbreviated as dbN, and N is the order number. Below is the data and SNR before and after the 2nd order Daubechies wavelet filtering. The green curve in the figure is the pre-filter curve, and the black and red are the filtered curves (Figs. 3 and 4). The ordinate is strength, and the abscissa is wavenumber. Figure 6 is the filtered noise.

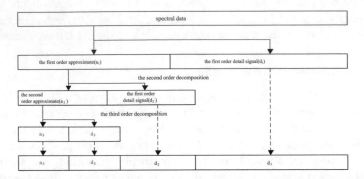

Fig. 3. Sketch map of multi-scale wavelet decomposition and construction of wavelet coefficient vector matrix

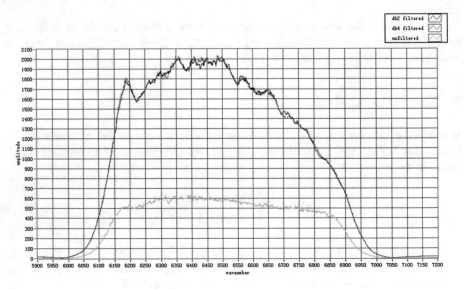

Fig. 4. The SNR before and after filtering

It can be seen from Figs. 5 and 6, the wavelet transform effectively filters out the noise.

3.1.2 Symlet Wavelet

Symlet wavelet is an approximate wavelet function proposed by Daubechies. It is an improvement of Db function. The Symlet function system is usually expressed as symN (N = 2, 3,... 8).

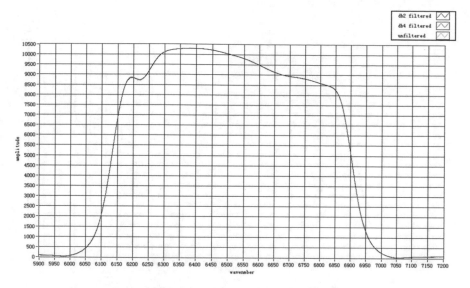

Fig. 5. The spectrum data before and after filtering

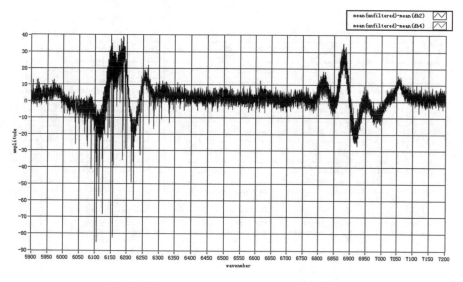

Fig. 6. The mean contrast before and after filtering

We can see from the figure below, useful information is reduced as the order increases (Figs. 7, 8 and 9).

3.1.3 Biorthorgonal Wavelet

In order to solve the incompatibility between symmetry and accurate signal reconstruction, biorthogonal wavelet is introduced. Two wavelets, called dual, are used for

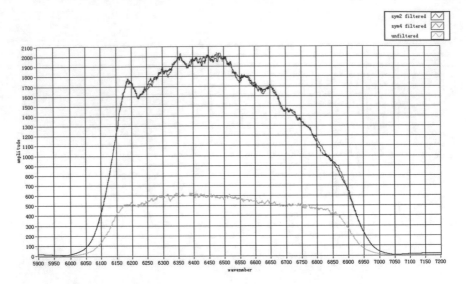

Fig. 7. The SNR before and after filtering

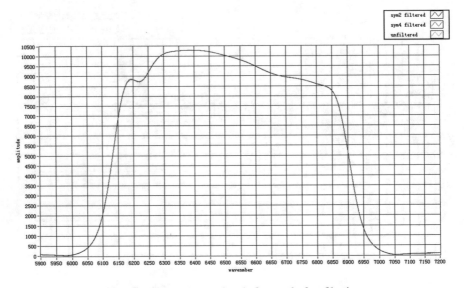

Fig. 8. The spectrum data before and after filtering

signal decomposition and reconstruction respectively. Biorthogonal wavelet solves the contradiction between linear phase and orthogonality. The main feature of Biorthorgonal function system is linear phase, so it is mainly used in the reconstruction of signal and image. The Biorthorgonal function is usually represented as a form of biorNr.Nd (R represents refactoring, and D means decomposition) (Figs. 10, 11 and 12).

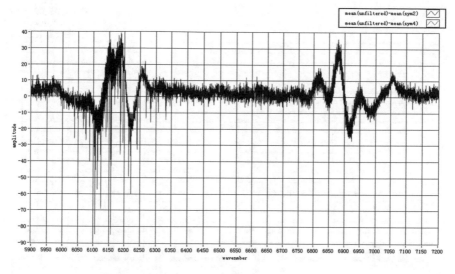

Fig. 9. The mean contrast before and after filtering

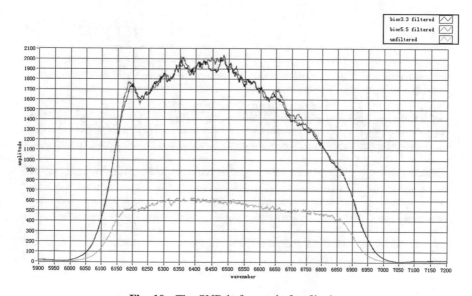

Fig. 10. The SNR before and after filtering

3.2 Comparison of Three Kinds of Wavelet Functions for Noise Reduction

Below is the SNR figure including three kinds of wavelet function(DB4, sym4, bior5.5) on the same data after filtering, and mean comparison of three kinds of wavelet function filter. (Sym4 wavelet denoising, the spectral mean minus the mean of DB4

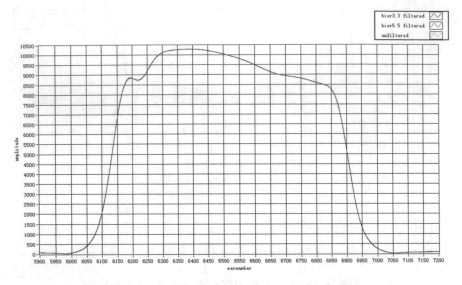

Fig. 11. The spectrum data before and after filtering

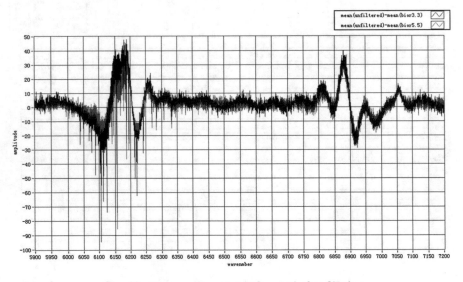

Fig. 12. The mean contrast before and after filtering

wavelet denoising, and sym4 wavelet denoising spectral mean minus the mean of bior5.5 wavelet denoising) (Figs. 13 and 14).

As we can see from the figure above, there are still differences among the three methods of reducing noise. The basic goal of application of wavelet denoising in the original spectral signal is not only to remove noise possibly, but hope the original signal loss as little as possible. While the two are not unified, sometimes even

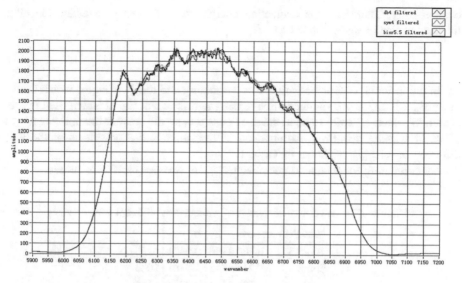

Fig. 13. The SNG figure of three filters

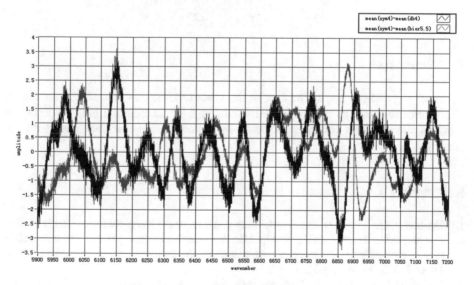

Fig. 14. The mean contrast of three filters

contradictory. Therefore, the evaluation coefficient η is used to evaluate the effectiveness of denoising. η defined as:

$$\eta = \sigma 12\sigma 22 \tag{3}$$

In the formula, $\sigma 12$ is the variance of de-noising and the ideal signal which used in noisy signal. $\sigma 22$ is the variance of de-noising and the ideal signal which used in ideal signal. We want $\sigma 12\sigma 22$ to be as small as possible, so the smaller the η, the better the effect of wavelet denoising. Comparison of the noise reduction effects of the three different wavelets is as shown in Table 1.

By comparing the evaluation coefficients η, he wavelet bior5.5 is chosen.

Table 1. The wavelet coefficients are evaluated by different wavelet basis

Wavelet bior	σ_1^2	σ_2^2	η
Bior5.5	704.4356	0.0801	56
Bior3	714.3345	1.5161	1083
sym4	693.0113	0.2353	163
db3	714.3345	1.5161	1083
db4	799.1358	0.4702	376
db5	789.5698	0.2235	176

4 Conclusion

By analyzing and processing the data of the spectrometer, it is proved that the digital filter is effective, and has the characteristics of high realization and high flexibility. It provides the direction for the application of hyperspectral observation load on the satellite.

References

1. Tianshuang, Q., Ying, G.: Signal processing and data analysis, pp. 142–146 (2015)
2. Edwards, G.C., Thurtell, G.W., Kidd, G.E.: A diode laser based gas monitor suitable for measurement of trace gas exchange using micro meteorological techniques. Agric. Forest Meteorol. **115**, 71–89 (2003)
3. Cassidy, D.T., Bonllell, L.J.: Trace gas detection with short-external-cavity InGaASP diode laser transmitter modules operating at 1.58 μm. Appl. Opt. **27**, 2688–2693 (1988)
4. Cancio, P., Corsi, C., Pavone, F.S., Martinelli, R.U., Menna, R.J.: Sensitive detection of ammonia absorption by using a 1.65 μm distributed feedback InGaAsP diode laser. Infrared Phys. Technol. **36**, 987–993 (1995)
5. Sonnenfroh, D.M., Allen, M.G.: Ultrasensitive visible tunable diode laser detection of NO_2. Appl. Opt. **35**, 4053–4058 (1996)

Identifying the Characteristic Scale of Satellite Images with Scale Space Representation and Normalized Derivatives

Xichao Teng[✉], Qifeng Yu, Zhihui Lei, and Xiaolin Liu

National University of Defense Technology, Changsha, China
tengari@buaa.edu.cn

Abstract. This paper presents an approach that identifies the characteristic scale of satellite images. It consists of three steps: (1) calculating the scale space representation of satellite images using convolution with the Gaussian kernel; (2) determining the local characteristic scale of every pixel by a local extremum over scale of normalized derivatives; (3) deriving the local characteristic scale histogram and determining the characteristic scale of satellite image based on maximum of this histogram. The local characteristic scale histogram reflects the local texture and geometric features while the characteristic scale represents the global salient structures of satellite images. Experiments on real data show that our proposed characteristic scale accurately identifies global structure of satellite images. We use the normalized Laplacian operator to determine the local characteristic in our experiments. Our proposed characteristic scale shed light on the structure understanding of satellite images. Moreover, the proposed local characteristic scale histogram can be extended to represent image structure for further scene classification of remote sensing images.

Keywords: Characteristic scale · Scale space · Normalized derivative
Local characteristic scale histogram · Salient structure · Satellite image

1 Introduction

A great deal of features have been proposed to describe the content of optical remote sensing images, and the feature extraction for remote sensing images is the foundations of many applications, such as image registration, image classification and image search. Among these features, many of them is related to the scale of remote sensing imagery, which can reveal the spatial distribution and resolution of remote sensing images. As scale is one of remarkable features for images, finding the characteristic scale of the image efficiently and accurately is of great importance. With the increase in sensor's spatial resolution, large number of details of image structure emerge from uniform image regions, and the scale of remote sensing images can characterize this process of change. In the case of optical satellite images, identifying the characteristic scale is useful for indexing the large and diverse satellite image database and image classification.

A lot of literature has been devoted to the representation of image scales. The popular scale space theory [1] was proposed by Lindeberg. Based on the scale space

© Springer Nature Switzerland AG 2018
H. P. Urbach and Q. Yu (Eds.): ISSOIA 2017, SPPHY 209, pp. 227–232, 2018.
https://doi.org/10.1007/978-3-319-96707-3_25

theory, many widely used methods for the scale space representation have been proposed, such as wavelet decompositions [2], pyramid decompositions [3], and scale invariant feature extraction [4]. Linderbeg defined the characteristic scale of local image regions by the local maximum of derivatives in [5], but for the characteristic scale of the whole image, [5] did not give a clear definition and the characteristic scale of satellite images have been little exploited. Chen et al. [6] used statistics of sub-images produced by a wavelet transform to identify the characteristic scale of satellite images. Luo et al. [7] used a linear scale space and the total variation to find the critical scale and the derived characteristic scale of satellite images was not depend on the spatial resolution.

In this paper, we assume the characteristic scale of optical satellite images is closely related to the physical size of most salient structure in remote sensing images, which resembles the definition of [7]. But our algorithm for extracting the characteristic scale can reflect more detail than the method used in [7]. Our algorithm first establishes the scale space representation of the satellite image. Then a normalized derivative is performed on the sub-regions of the image to evaluate the local scale. The characteristic scale of sub-images is defined as the one at which the normalized derivative reaches its local maximum. Finally, our method counts the number of pixels for every scale and get a characteristic scale histogram. The characteristic scale for the whole image is defined as the one at which the characteristic scale histogram reaches its maximum. Compared to [7] which is an average measure, our characteristic scale histogram can reflect features of the satellite images at different scales as it is the statistics produced by local characteristic scale at every pixel and we can see the distribution of different scales in the image. For thinner or coarser scales than the characteristic scale, the characteristic scale histogram can also imply the structure feature of corresponding sub-regions.

Our paper is organized as follows. In Sect. 2, we describe our algorithm for finding the characteristic scale in detail. In Sect. 3, we test our method on satellite images and discuss the results. Finally, Sect. 4 analyzes the method and concludes the paper.

2   Methods

Our method for determining the characteristic scale of the image includes three steps:

1. Calculating the scale space representation of the image using convolution with the Gaussian kernel;
2. Finding the local characteristic scale of every pixel by a local extremum over scale of normalized derivatives;
3. Deriving the characteristic scale histogram using the results of the local characteristic scale and identifying the characteristic scale of the whole image based on maximum of this histogram.

In this section, we will introduce our algorithm in this three parts.

2.1 Scale Space Representation

The scale space theory is a basic mathematics tool to analyze structure at different scales. The scale space representation can be performed on signals of arbitrary numbers of variables. The most common case of scale space applies to two dimensional images, by representing an image as a group of smoothed images over scale parameter. The main type of scale space is the Gaussian scale space, given a scale parameter s, the Gaussian scale space representation $L(\mathbf{x}, s)$ is denoted as following:

$$L(\mathbf{x}, s) = G(\mathbf{x}; s) * I(\mathbf{x}) = \frac{1}{2\pi s^2} e^{-\frac{|\mathbf{x}|^2}{2s^2}} * I(\mathbf{x}) \tag{1}$$

Where $I(\mathbf{x})$ refer to the image, $G(\mathbf{x}; s)$ is the Gaussian kernel with mean zero and standard deviation s, and $*$ refer to linear convolution. The scale parameter is the standard deviation s of the Gaussian filter and typically only a finite discrete set of s in the scale space representation would be actually considered. We assume that s_0 is the initial scale factor at the finest level of resolution, then successive levels of the scale space representation s_n is denoted by $s_n = k^n s_0$, where k is the factor of scale change between successive levels.

Figure 1 illustrates the scale space representation of a satellite image. As we can see, the image is more and more smooth with the increase of scale parameter s.

2.2 Normalized Derivatives

We can extract features related to scale by applying a derivative in the scale space. In general, the amplitude of the derivative over scales will decrease with the increase in levels which means that the derivative is not scale invariant. In order to achieve the scale invariance, we normalize the derivative function with respect to the scale parameter. The normalized derivative $F(\mathbf{x}; s)$ of order m is denoted by:

$$F_{(m)}(\mathbf{x}; s) = s^m L_{(m)}(\mathbf{x}; s) = s^m G_{(m)}(\mathbf{x}; s) * I(\mathbf{x}) \tag{2}$$

The properties of several normalized derivatives were studied in [8] and the experimental results in [8] showed that for a single function Laplacian achieved the best performance for finding the correct local characteristic scale. In our paper, normalized Laplacian is used to compute a scale representation of a satellite image, which is denoted as follows:

$$F_{(2)}(\mathbf{x}; s) = s^2 |L_{xx}(\mathbf{x}; s) + L_{yy}(\mathbf{x}; s)| \tag{3}$$

We can get the local characteristic scale in a sub region around a pixel by searching the local maximum of normalized derivative over scales. For a given pixel in an image, the amplitude of normalized Laplacian is calculated at different scales, as shown in Fig. 2. If there are several local maxima, several local characteristic scales will be determined.

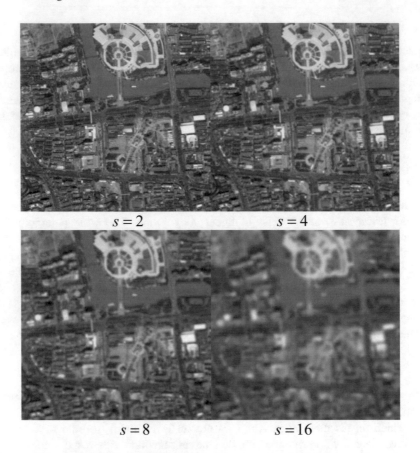

$s = 2$ $s = 4$

$s = 8$ $s = 16$

Fig. 1. The scale space representation of a satellite image

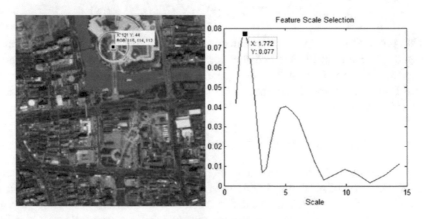

Fig. 2. The responses of normalized Laplacian for a given pixel

2.3   Characteristic Scale Histogram

We select the local characteristic scale for every pixel in the image by searching the local maxima of the amplitude of normalized Laplacian. After that, we get a new two dimensional image in which the value of each pixel refer to the local characteristic scale of its neighborhood. In order to identify the characteristic scale of the whole image, we calculate the statistics of this new scale image with the information of local characteristic scales. Our algorithm calculates the histogram of the scale image which we call the characteristic scale histogram and the maximum of the characteristic scale histogram is defined as characteristic scale of the whole image.

3   Experiments

We use satellite images in different areas to test our algorithm. A satellite image in desert area with the size of 512 × 512 pixels (spatial resolution 1.6 m) is shown in Fig. 3. A satellite image in urban area with the size of 512 × 512 pixels (spatial resolution 1.6 m) is shown in Fig. 4.

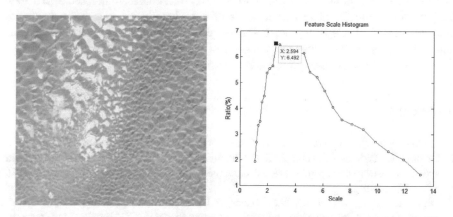

Fig. 3. The characteristic scale histogram of a satellite image in desert area

Figure 3 shows the characteristic scale histogram of a satellite image in desert area. The scale at which the histogram reaches its maximum is the characteristic scale of the satellite image.

Figure 4 shows the characteristic scale histogram of a satellite image in urban area. The characteristic scale is marked on the histogram. Compared with the characteristic scale of Fig. 3, the one of Fig. 4 is larger and this is consistent with our observation of this two satellite images. The size of salient structure in the satellite image of Fig. 3 is smaller than the size in the satellite image of Fig. 4. The characteristic scale reflects the dimension of most salient structure in the scene.

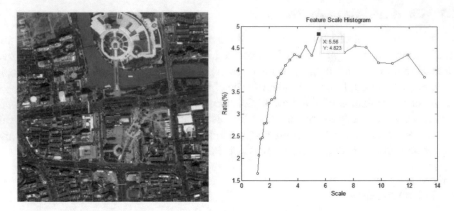

Fig. 4. The characteristic scale histogram of a satellite image in urban area

4 Conclusion

In this paper, we propose a method to identify the characteristic scale of satellite images. For finding the characteristic scale of the whole image, our method calculates the characteristic scale histogram based on the scale space representation and normalized derivatives. The characteristic scale of the satellite image is the scale at which the characteristic scale histogram reaches its maximum.

References

1. Lindeberg, T.: Scale-Space Theory in Computer Vision. Kluwer, Dordrecht (1994)
2. Demirel, H., Ozcinar, C., Anbarjafari, G.: Satellite image contrast enhancement using discrete wavelet transform and singular value decomposition. IEEE Geosci. Remote Sens. Lett. **7**(2), 333–337 (2010)
3. Montoya-Zegarra, J.A., Leite, N.J., Torres, R.D.S.: Rotation-invariant and scale-invariant steerable pyramid decomposition for texture image retrieval. In: XX Brazilian Symposium on Computer Graphics and Image Processing, pp. 121–128. IEEE Computer Society (2007)
4. Lowe, D.G.: Distinctive Image Features from Scale-Invariant Keypoints. Kluwer, Dordrecht (2004)
5. Lindeberg, T.: Feature detection with automatic scale selection. Int. J. Comput. Vis. **30**, 79–116 (1998)
6. Chen, K., Blong, R.: Identifying the characteristic scale of scene variation in fine spatial resolution imagery with wavelet transform-based sub-image statistics. J. Shanghai Jiaotong Univ. **24**(9), 1983–1989 (2007)
7. Luo, B., et al.: Resolution-independent characteristic scale dedicated to satellite images. IEEE Trans. Image Process. **16**(10), 2503–2514 (2007)
8. Mikolajczyk, K., Schmid, C.: Indexing based on scale invariant interest points. In: Proceedings of Eighth IEEE International Conference on Computer Vision, ICCV 2001, vol. 1, pp. 525–531. IEEE (2002)

An Efficient Cloud Detection Algorithm Based on CNN and SVM with Its Prospect in Aerospace Camera

Zhicheng Yu[✉], Bingxin Yang, and Tao Li

Key Laboratory for Advanced Optical Remote Sensing Technology of Beijing,
Beijing Institute of Space Mechanics and Electricity, Beijing, China
23326026@qq.com

Abstract. Terrain information is difficult to obtain due to cloud cover, these regions will become blind in remote sensing images, which led to the remote sensing images can not play its due value. So we need to use some detection technology to detect the cloud in remote sensing image to avoid a lot of useless images captured by the satellite wastes a large amount of storage space and transmission bandwidth. This paper relay on a delicate high-resolution earth observation satellite in synchronous orbit, combined the CNN with SVM for cloud detection, and finally designed a new algorithm which can be realized in remote sensing satellite video processing system to statistic the cloud coverage ratio in orbit. Firstly, the large image is cut into small pieces, which will be send into the algorithm module. The new algorithm uses the deep learning algorithm to automatically abstract the effective features, and after that SVM classifier is used to recognize the thick cloud. The CNN net is designed with a matlab toolbox Matconvnet, which is easy to use and provides many pre-trained model. The most important is that the CNN network is simple enough so that it can be realized in embedded system like Zynq platform. Final 1000 cloud subimages and 1000 no-cloud subimages are used as training dataset, and another 1000 cloud subimages and 1000 no-cloud subimages are used as testing dataset, and the illation labels are compared to the true labels, the result shows that this algorithm is much more accurate than traditional cloud detection such as threshold method or SVM. In the future, this new algorithm can be easily combined with dynamic partial reconfiguration of FPGA, then the algorithm codes can be upload from earth to the satellite and reconfigure the algorithm module in FPGA. So the satellite can advance with the times.

Keywords: Cloud detection · Deep learning · SVM · CNN

1 Introduction

1.1 Definition of Cloud Detection

Cloud detection, also known as cloud segmentation and cloud filtering, is to determine the cloud content in satellite remote sensing images. Associated with another concept, known as the cloud classification or recognition of cloud, it is to have determined the cloud of a region known as few classes (cumulus clouds cirrus clouds, such as clouds,

© Springer Nature Switzerland AG 2018
H. P. Urbach and Q. Yu (Eds.): ISSOIA 2017, SPPHY 209, pp. 233–242, 2018.
https://doi.org/10.1007/978-3-319-96707-3_26

etc.). Together they can be called cloud identification. Cloud detection can be seen as a binary partition task for image analysis, while cloud recognition is a typical classification task. To avoid confusion, only the term for cloud detection is applicable in future text. This paper studies the problem of cloud detection.

1.2   The Present Situation of Remote Sensing Image Cloud Pollution

Remote sensing technology has played an increasingly important role because of its quick, timely, convenient to provide a variety of earth observation, tracking, positioning and navigation and other services, and bring a huge military benefit, social benefit and economic benefit in many aspects, such as national defense modernization and economic construction. However, due to the presence of cloud cover, the remote sensing image cannot be used for the purpose. And because of the effect of cloud cover, a remote sensing image will not be able to obtain the earth surface or other information that we need. And because of the climate, it's hard to get remote sensing images with no cloud noise pollution, so cloud is an important problem to be solved in remote sensing images, and how to find an effective method to recognize the cloud is a very important research topic in the field of remote sensing image processing.

Statistics show that over 65% of remote sensing images are cloud or cloudy. Especially collected in visible light remote sensing satellite remote sensing images are more or less pollution clouds, rarely collected clean image, according to data shows, the Landsat 7 ETM+ were collected from 2000 to 2002, the global annual average cloud cover for remote sensing images, cloud coverage in more than 33%.

From 2000 to 2000, even though the weather forecast and automatic cloud cover algorithm ACCA, improve the efficiency of image acquisition, the U.S. bureau of surveying and mapping (USGS/EROS) archived Landsat 7 remote sensing image, the cloud cover over 10% of accounts for more than 65% of the entire image; Accordingly, the amount of cloud above 50% of the total image is over 28% (Table 1).

Table 1. Global annual average cloud cover situation

	Annual	Winter	Spring	Summer	Autumn
2000	0.36 (0.33)	0.44 (0.31)	0.33 (0.30)	0.26 (0.35)	0.42 (0.37)
2001	0.43 (0.36)	0.54 (0.33)	0.45 (0.34)	0.34 (0.39)	0.37 (0.38)
2002	0.42 (0.35)	0.49 (0.35)	0.45 (0.33)	0.31 (0.37)	0.44 (0.35)

Because cloud pattern, brightness, texture is always over time, height, thickness, Angle of the sun, and other factors changes, the cloud detection is one of the most difficult part of the remote sensing image processing.

At home and abroad in recent years, the research on cloud detection in orbit enlarges gradually, but there is still no particularly effective project implementation experience. The main difficulties are as follows:

(1) Most traditional cloud detection technology is based on threshold method to judge whether it is covered by cloud, this method is highly vulnerable to light, the zenith

Angle, shooting Angle and ground objects misjudgment caused by the influence of, and the need for plenty of remote sensing data statistics, to draw more accurate threshold;

(2) Traditional pattern recognition method such as support vector machine (SVM) need manual specify the classification features, the types of clouds is very complex, highly, texture is different, the shape and protean, unable to effectively extract the projections for edge character, it is difficult to find effective characteristics, extremely hard detection;

(3) It is especially easy to mix with the ice and snow on the ground, adding to the difficulty of detecting it.

(4) The precision of theoretical research is conducted in multispectral data feature extraction, but this solution cannot be achieved in orbit for on-board resources are limited, storage capacity and the star on the processor speed cannot meet the requirements;

In this paper, I find an effective cloud detection algorithm, and in the case of on-board resources allow, achieve ideal cloud detection accuracy.

2 Computational Blocks in MatConvNet

MatConvNet is an implementation of Convolutional Neural Networks (CNNs) for MATLAB. The toolbox is designed with an emphasis on simplicity and flexibility. It exposes the building blocks of CNNs as easy-to-use MATLAB functions, providing routines for computing linear convolutions with filter banks, feature pooling, and many more. In this manner, MatConvNet allows fast prototyping of new CNN architectures; at the same time, it supports efficient computation on CPU and GPU allowing to train complex models on large datasets such as ImageNet ILSVRC.

2.1 Convolution

The convolutional block is implemented by the function vl_nnconv. y = vl_nnconv (x, f, b) computes the convolution of the input map x with a bank of K multi-dimensional filters f and biases b.

The process of convolving a signal is illustrated in Fig. 1 for a 1D slice. Formally, the output is given by

$$y_{i''j''d''} = b_{d''} + \sum_{i'=1}^{H'} \sum_{j'=1}^{W'} \sum_{d'=1}^{D} f_{i'j'd} \times x_{i''+i'-1,j''+j'-1,d',d''}.$$

The call vl_nnconv(x,f,[]) does not use the biases. Note that the function works with arbi- trarily sized inputs and filters (as opposed to, for example, square images).

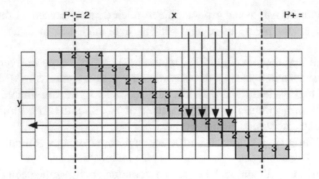

Fig. 1. Convolution process diagram

2.2 Spatial Pooling

vl_nnpool implements max and sum pooling. The max pooling operator computes the max- imum response of each feature channel in a $H' \times W'$ patch

$$y_{i''j''d} = \max_{1 \le i' \le H', 1 \le j' \le W'} x_{i''+i'-1, j''+j'-1, d}$$

resulting in an output of size $y \in RH'' \times W'' \times D$, Sum-pooling computes the average of the values instead:

$$y_{i''j''d} = \frac{1}{W'H'} \sum_{1 \le i' \le H', 1 \le j' \le W'} x_{i''+i'-1, j''+j'-1, d}$$

2.3 Activation Functions

MatConvNet supports the following activation functions:

(1) ReLU
   vl_nnrelu computes the Rectified Linear Unit (ReLU):

$$yijd = \max\{0, \ xijd\}$$

(2) Sigmoid
   vl_nnsigmoid computes the sigmoid:

$$y_{ijd} = \sigma(x_{ijd}) = \frac{1}{1+e^{-x_{ijd}}}.$$

2.4 Batch Normalization

vl_nnbnorm implements batch normalization. Batch normalization is somewhat different from other neural network blocks in that it performs computation across images/feature maps in a batch (whereas most blocks process different images/feature maps individually). y = vl_nnbnorm(x, w, b) normalizes each channel of the feature map x averaging over spatial locations and batch instances.

Note that in this case the input and output arrays are explicitly treated as 4D tensors in order to work with a batch of feature maps. The tensors w and b define component-wise multiplicative and additive constants. The output feature map is given by

$$y_{ijkt} = w_k \frac{x_{ijkt} - \mu_k}{\sqrt{\sigma_k^2 + \epsilon}} + b_k, \quad \mu_k = \frac{1}{HWT} \sum_{i=1}^{H} \sum_{j=1}^{W} \sum_{t=1}^{T} x_{ijkt},$$

$$\sigma_k^2 = \frac{1}{HWT} \sum_{i=1}^{H} \sum_{j=1}^{W} \sum_{t=1}^{T} \left(x_{ijkt} - \mu_k \right)^2$$

2.5 Softmax

vl_nnsoftmax computes the softmax operator:

$$y_{ijk} = \frac{e^{x_{ijk}}}{\sum_{t=1}^{D} e^{x_{ijt}}}$$

Note that the operator is applied across feature channels and in a convolutional manner at all spatial locations. Softmax can be seen as the combination of an activation function (exponential) and a normalization operator.

3 Technical Protocols

This article combines CNN with SVM, using CNN to extract the eigenvectors and using SVM to classify features. By using convolution and pooling, avoiding the effect of translation on the final extraction eigenvectors, the extracted features are less likely to be matched; Secondly, using CNN to extract feature than simple grayscale, projection and texture, gray level should be more scientific, avoid the become the bottleneck of finally improve the accuracy of feature extraction; The size of the overall model can be controlled by different convolution, pooling, and final output eigenvectors. It can reduce the dimension of the eigenvector in the fitting, and it can improve the output dimension of the convolution layer at the time of the ill-fitting, which is more flexible than the other feature extraction methods. The algorithm process used in this paper is divided into two parts: the training process and the test procedure.

When training, firstly find a mature convolutional neural network pre-trained model, in this paper I use the VGG net, using training samples to optimize the network, get the optimized neural network, take out the classifier of the neural network, then replace the original classifier of the network with the SVM classifier, directly connected SVM with the feature vectors of network layer, input the training samples to the training model to get feature vectors, training the SVM model, finally getting the SVM model parameters, then, as shown in Fig. 2 of dotted box, the optimized network (CNN) and the trained SVM model together constitute the core of this algorithm.

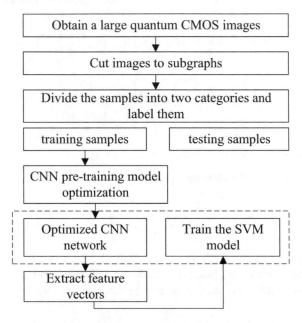

Fig. 2. Model parameter training flow

In the software, I firstly cut the large images to small pieces of 255 * 255 pixels, and classify them to none-cloud group and cloud group, and put them into different file folders. Then use the "imread" command to load the training images into a big variable "im2". I use the "net.meta.normalization.imagesSize" command regulation the im2 to "im_". Next, "im_" subtrace the average value and get the final "im_", load the vgg net to matlab and use the "vl_simplenn" command to train the net best CNN net model, drop the last layer of the net and save the sub-last layer result to the variable "train-data", then label these training images with a variable "trainlabel" filled with constant '1'. Next, I use the "svmtrain" command to get a SVM model with two main parameters: traindata and trainlabel. So the kernel function is rbf by default, rbf sigma is 1, polyorder is 3. When the svmtrain function is run over, I get a SVM model.

As shown in the Fig. 3, in test process, input the marked test samples to the optimized network (CNN), extraction of the feature vectors, and then pass the feature vectors to the SVM classifier for classification results of each test samples.

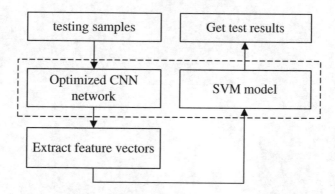

Fig. 3. Model parameter testing process

Similarly, in the software of the test process, use the "imread" command to load the test images into a big variable "im2", use the "net.meta.normalization.imagesSize" command regulation the im2 to "im_". Next, "im_" subtrace the average value and get the final "im_", load the vgg net to matlab and use the "vl_simplenn" command to get the sub-last layer result of CNN net as the feature vector to the variable "testdata", then input these test images to the SVM model with the "smvclassify" command to get a label as the classfy result.

4 Results and Analysis

In this paper, 500 cloud and 500 non-cloud subplots were selected as training samples, another 500 cloud and 500 non-cloud subplots as the testing samples. Finally, the result shows that the accuracy is 88%.

The cloud subplots samples selected for experiment is as follows (Fig. 4):

The non-cloud subplots samples selected for experiment is as follows (Fig. 5):

However, most misjudged subgraph are as followed (Fig. 6):

As is shown above, we can see that: firstly, most miscalculations didn't come from the accuracy of the algorithm itself, but rather from the quality of the images taken by the space camera itself. If the subgraph is good enough, the texture is clear and the exposure is modest, then the accuracy of the algorithm can be greatly improved; and the most important thing is that new algorithm uses the deep learning algorithm to automatically abstract the effective features, and after that SVM classifier is used to recognize the thick cloud, which avoid human choose the feature of the cloud, which is very difficult and easy to make a mistake.

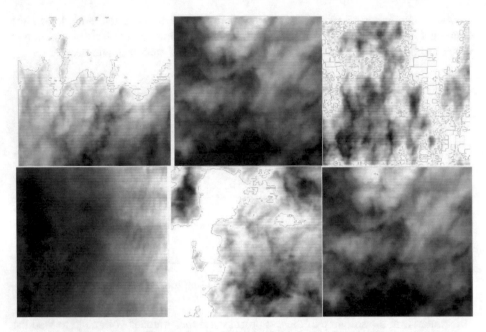

Fig. 4. Standard cloud subgraph samples

Fig. 5. Standard non-cloud subgraph samples

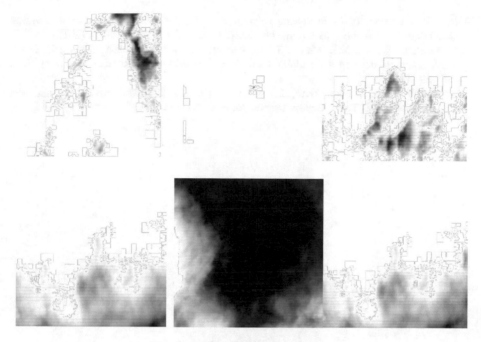

Fig. 6. Misjudged subgraph samples

References

Sarkar, N., Chaudhuri, B.B.: An efficient differential box-counting approach to compute fractal dimension of image. IEEE Trans. Syst. Man Cybern. **24**(1), 115–120 (1994)

Zhirong, L., Mcdonnell, M.J.: Using dual thresholds for cloud and mosaicking. Int. J. Remote Sens. **7**(10), 1349–1358 (1986)

Shin, D., Pollard, J.K.: Cloud detection from ATSR images using a segmentation techniques. SPIE **2578**, 46–52 (1995)

Saunders, R.W.: An automated scheme for the removal of cloud contamination from AVHRR radiances over western Europe. Int J. Remote Sens. **7**(7), 867–886 (1986)

Zavody, A.M., Mutlow, C.T., Llewelyn-Jones, D.T.: Cloud clearing over the ocean in the processing of data from the along-track scanning radiometer (ATSR). J. Atmos. Ocean. Technol. **17**(5), 595–615 (2000)

Liu, R., Liu, Y.: Generation of new cloud masks from MODIS land surface reflectance products. Remote Sens. Environ. **133**(15), 21–37 (2013)

Sedano, F., Kempeneers, P., Strobl, P., et al.: A cloud mask methodology for high resolution remote sensing data combining information from high and medium resolution optical sensors. ISPRS J. Photogramm. Remote Sens. **66**(5), 588–596 (2011)

Jedlovec, G.J., Haines, S.L., Lafontainef, J.: Spatial and temporal varying thresholds for cloud detection in GOES imagery. IEEE Trans. Geosci. Remote Sens. **46**(6), 1705–1717 (2008)

Luo, Y., Trishchenko, A.P., Khlopenkov, K.V.: Developing clearsky, cloud and cloud shadow mask for producing clear-sky composites at 250-meter spatial resolution for the seven MODIS land bands over Canada and North America. Remote Sens. Environ. **112**(12), 4167–4185 (2008)

Wilson, M.J., Oreopoulos, L.: Enhancing a simple MODIS cloud mask algorithm for the Landsat data continuity mission. IEEE Trans. Geosci. Remote Sens. **51**(2), 723–731 (2013)

Oreopoulos, L., Wilson, M.J., Vhrnai, T.: Implementation on Landsat data of a simple cloud-mask algorithm developed for MODIS land bands. IEEE Geosci. Remote Sens. Lett. **8**(4), 597–601 (2011)

Leiva-Murillo, J.M., Gomez-Chova, L., Camps-Valls, G.: Multitask remote sensing data classification. IEEE Trans. Geosci. Remote Sens. **51**(1), 151–161 (2013)

Design and Manufacture of the Compact Conical Diffraction Imaging Spectrometer

Qiao Pan[1,2], Xinhua Chen[1,2(✉)], Zhicheng Zhao[1,2], Quan Liu[1,2], Xiaofeng Wang[1,2], Chao Luo[1,2], and Weimin Shen[1,2]

[1] College of Physics, Optoelectronics and Energy,
Soochow University, Suzhou 215006, China
`xinhua_chen@suda.edu.cn`
[2] Key Lab of Advanced Optical Manufacturing Technologies of Jiangsu Province and Key Lab of Modern Optical Technologies of Education Ministry of China, Soochow University, Suzhou 215006, China

Abstract. The conical diffraction imaging spectrometer, which is based on the off-plane Offner mounting, can achieve high spectral resolution, low-level distortion with compact package. It has attracted much attentions recently and been used for many fields in which the dispersion width is much longer than the slit length. The design and manufacture of a compact conical diffraction imaging spectrometer, which operates at F/4 and covers the 400–900 nm spectral region, is introduced is this paper. The body of the imaging spectrometer is made of titanium alloy and can work stably in a wide temperature range. The MTF and the spectral performance of the imaging spectrometer are measured after alignment. MTF measurement shows that the imaging quality of this imaging spectrometer is almost diffraction limited. The measured spectral resolution is 4.8 nm, while the smile and keystone is about 4% and 3.4% of a pixel width respectively.

Keywords: Conical diffraction imaging spectrometer · Convex blazed grating
Spectral resolution

1 Introduction

The Offner imaging spectrometers using the convex grating is preferred for spectral imaging applications. It can realize low F-number, excellent imaging quality with low smile and keystone, and has been used widely in remote sensing, target recognition, ocean monitoring and so on [1, 2]. It develops from the concentric and symmetrical Offner configuration proposed by Offner [3]. After that, Das suggested using a convex holographic grating to take place of the convex mirror of the Offner relay, and this is the so-called in-plane Offner mounting [4]. In this configuration, the entrance slit and the grating groove orientations are both perpendicular to its optical symmetric plane. Usually, it is very suitable for a long entrance slit and narrow dispersion width. However, another off-plane configuration is more attractive when a short entrance slit but a wide dispersion width is required. In this configuration, the silt and the spectral image are on the opposite sides of the diffraction grating in a direction perpendicular to

© Springer Nature Switzerland AG 2018
H. P. Urbach and Q. Yu (Eds.): ISSOIA 2017, SPPHY 209, pp. 243–253, 2018.
https://doi.org/10.1007/978-3-319-96707-3_27

the slit itself [5–8]. Compared with the in-plane configuration, the grating works in the conical diffraction mode in the off-plane configuration.

In this paper, we introduce the design and manufacture of a conical diffraction imaging spectrometer, which is actually based on the off-plane Offner mounting. The spectral range of this spectrometer is from 400 to 900 nm. Its entrance slit length is 1 mm, while the dispersion width is 10.4 mm, which is about ten times longer than the slit length. The optical design of the imaging spectrometer is introduced in Sect. 2. The grating manufacture and the alignment of the imaging spectrometer are introduced in Sect. 3. The measurement of the modulation transfer function (MTF) and the spectral performance of the imaging spectrometer are presented in the Sect. 4. In Sect. 5, we give conclusions of the design and manufacture of the imaging spectrometer.

2 Optical Design

2.1 Conical Diffraction Grating

In the off-plane configuration, the grating works in the conical diffraction mode as shown in Fig. 1 [9]. The origin of the coordinate is placed at the grating vertex O. The Y axis is the normal to the grating plane at this point and the grating grooves is parallel to the Z axis with regular spacing d. The direction vectors of incident and diffractive light are defined as $\bar{n}$ and $\bar{n}'$, respectively. The plane perpendicular to the grating groove, i.e. the XY plane in Fig. 1, is defined as the principle section of the grating. According to the grating equation and the geometric relationships, we can get the following equation:

$$cos\,\alpha\,sin\,\theta_i + cos\,\alpha'\,sin\,\theta_K = \frac{K\lambda}{d} \qquad (1)$$

where K is spectral order and λ is wavelength, α/α' describes the orientation of the plane of incidence/diffraction with respect to the XY plane while angle θ_i/θ_K stands for the angle between the projection of $\bar{n}/\bar{n}'$ in the principal section of the grating and positive Y axis. This equation can be used in the calculation of the initial structure parameters of the imaging spectrometer.

2.2 Optical Design Result

The typical optical performance requirements are given in the Table 1. We can find that the required slit length is 1 mm, while the dispersion width is 10.4 mm. The off-plane Offner mounting is an excellent choice for this condition. The required silt width and the detector pixel size are as large as 104 μm, so a simple reflective system consisting of one or two mirrors can be used as the objective of the imaging spectrometer. We finally use a single off-axis parabolic mirror as the objective. Its focal length is 74.4 mm, and the off-axis distance is 50 mm to leave enough space for the mechanical structure. The Offner mounting consist of two spherical mirrors and a reflective convex blazed grating. The whole imaging spectrometer is designed with the commercial optical design software ZEMAX, and the final optical layout is shown in Fig. 2.

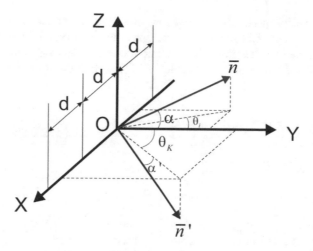

Fig. 1. Diagram of a spherical grating

Table 1. Optical parameter of the imaging spectrometer

Parameters	Values
Spectral range	400–900 nm
F-number	4
Focal length	74.4 mm
Slit length/width	1 mm/104 μm
Dispersion width	10.4 mm
Spectral resolution	5 nm
Smile	≤ 1/10 pixel
Distortion	≤ 1/10 pixel

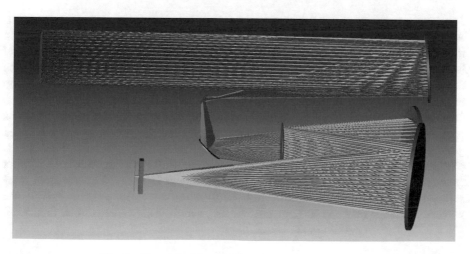

Fig. 2. The optical layout of the imaging spectrometer

For the detector pixel size is 104 μm, the Nyquist frequency of the imaging spectrometer is about 4.8l p/mm. The designed MTF curves of the imaging spectrometer when the wavelength is 400 nm and 900 nm respectively are given in Fig. 3. We can find that the imaging performance is almost diffraction limited at the Nyquist frequency.

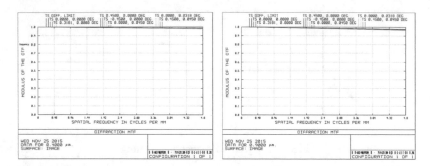

Fig. 3. The designed MTF of the imaging spectrometer at 400 nm and 900 nm

The smile and the keystone are the two important parameters for the imaging spectrometer. The smile describes the distortion for different fields, while the keystone describes the magnification change for different wavelengths. They can be evaluated by calculating the spot image centroids for different fields and different wavelengths, and the designed results are shown in Fig. 4.

3   Manufacture and Alignment of the Imaging Spectrometer

3.1   Manufacture of the Convex Blazed Grating

The convex grating is the key component in the Offner imaging spectrometer. According to the optical design result, the period of the grating is 2.45 μm. In order to improve the first-order diffraction efficiency, the groove profile of the grating should have a sawtooth profile. The blaze angle of the grating is 6.9° after optimized with the Comsol software. The grating is manufactured by holographic lithography - scan ion beam etching method on a convex substrate and the final grating is shown in Fig. 5 [10]. The groove profile of the grating is measured with the Bruker Dektak XT stylus profiler, and the result is given in Fig. 6. It shows that the period of the fabricated grating is 2.45 μm, which is equal to the required value. The diffraction efficiency is also measured by using a similar Offner mounting and a coating substrate without grating structure, and the measurement result is shown in Fig. 7. Experimental measurement shows that the first-order diffraction efficiency is more than 45% in the wavelength range between 400 nm and 900 nm, and the max value is more than 85%.

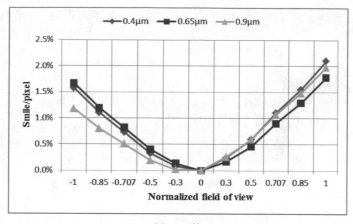

(a) smile

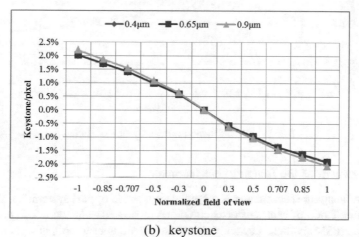

(b) keystone

Fig. 4. The designed (a) smile and the (b) keystone of the imaging spectrometer

Fig. 5. The fabricated convex blazed grating

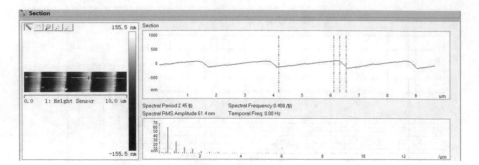

Fig. 6. The measured groove profile of the grating

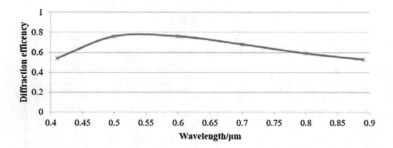

Fig. 7. The diffraction efficiency of the grating

3.2   Alignment of the Imaging Spectrometer

The Offner imaging spectrometer is a typical concentric optical system, i.e. the centers of curvature (CoC) of each mirror is coincided or almost coincided. The point source microscope (PSM), which is actually an autostigmatic microscope, is a good choice to align this type system. The PSM mainly consist of a microscope objective, a diode laser, a pinhole and a CCD camera [11]. It can be used to find the CoCs of the lenses or mirrors during the alignment, and this is very helpful for the concentric system alignment. We first place a steel ball on the designed place marked on the imaging spectrometer, the center of the ball is coincided with the designed CoC of the primary mirror by measuring with the coordinate measurement machine. We move the PSM until it focus on the center of the ball and this means that the spot source generated by the PSM is coincide with the center of the ball. We set the position of the spot image captured by the CCD camera as the reference position in the control software of PSM. Then we move away the steel ball and adjust the primary mirror (PM) and the tertiary mirror (TM) until their reflection images are coincide with the reference point. For the grating surface is convex, another concave spherical mirror is used to help the alignment. The concave mirror is attached on the back of the grating, and the thickness between it and the grating is designed to ensure its COC coinciding with the PM and TM's when the grating is perfectly assembled. After that, we rotate the grating slightly until the measured smile and keystone of the imaging spectrometer meet the requirement. During the rotation, we should confirm that there is no movement of the reflection image of the concave mirror. Finally, we move away the concave mirror and complete the alignment of the imaging spectrometer. The optical layout during the alignment of the imaging spectrometer is shown in Fig. 8 and the photo of the PSM and imaging spectrometer is given in Fig. 9.

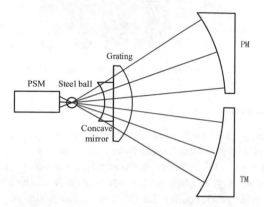

Fig. 8. The optical layout during the alignment of the imaging spectrometer

Fig. 9. The PSM and the imaging spectrometer during the alignment

4 Performance Measurements

The final imaging spectrometer after assembly and alignment is presented in Fig. 10. Its body is made of the titanium alloy to adapt to the change of environment temperature. The whole size of the imaging spectrometer is only 113.03 × 110.37 × 85.75 mm, and the weight is 987.5 g. In order to evaluate its performance, we measure the MTF and the spectral performance and the results are given in the following section.

Fig. 10. The photo of the final imaging spectrometer

4.1 MTF Measurement

MTF is used widely to evaluate the imaging performance of optical systems. It describes how image contrast varies as the spatial frequency increases. Several methods can be used to measure MTF by using different target such as gratings, pinholes, slits, knife edge and random targets. Here, we measure the MTF of the imaging spectrometer with the Optest system produced by the Optikos corporation. A pinhole with 23 μm diameter is place at the focus of the 3000 mm collimator, and the image analyzer of the Optest system is used to capture the spot images of the imaging spectrometer. We measure the MTF at the three wavelengths: 488 nm, 532 nm and 632 nm, and the measurement results are given in Fig. 11. The MTF of the imaging spectrometer are almost diffraction limited at the three wavelengths.

4.2 Spectral Performance Measurement

In this measurement, the spectral resolution, smile and keystone are measured with the captured spectral image. The Hg-Cd lamp is used as the source in the measurement and it can provide very narrow spectral line at several discrete wavelengths. We use the lamp to illuminate the imaging spectrometer directly, and a CCD camera with a pixel size of 4.54 μm is assembled at the imaging plane of the imaging spectrometer to capture the spectral images. The measurement instrument is shown in Fig. 12.

The intensity distribution profile of the middle row in the captured spectral image is shown in Fig. 13. The average linear dispersion can be calculated with the corresponding wavelength of each spectral line and the pixel size, and its value is about 0.0206 mm/nm. The spectral resolution can be measured by calculating the full width at half maximum (FWHM) of the spectral lines, and the average spectral resolution of this imaging spectrometer is about 4.8 nm.

The smile and keystone are also measured by calculating the centroids for different fields at the specified wavelengths corresponding to the characteristic spectral lines of Hg-Cd lamp. The maximum smile of this imaging spectrometer is 4% of pixel width, while the maximum keystone is only 3.4% of pixel width.

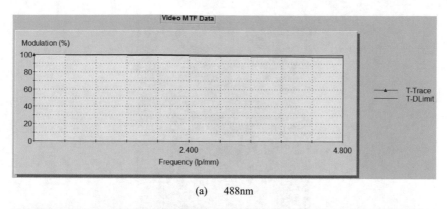

(a)   488nm

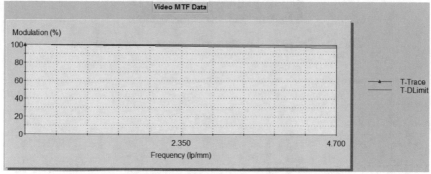

(b)   532nm

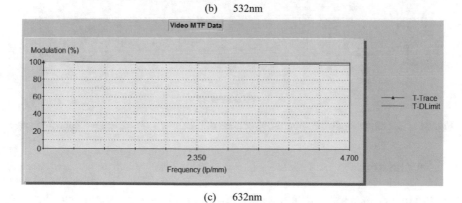

(c)   632nm

Fig. 11. The measured MTF of the imaging spectrometer

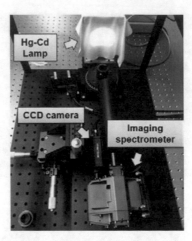

Fig. 12. The spectral performance measurement instrument

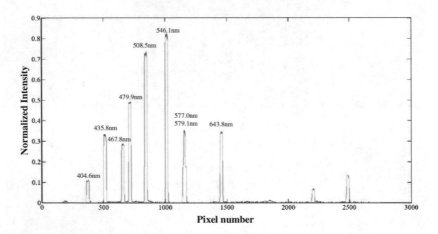

Fig. 13. The intensity distribution profile of the middle row in the captured spectral image

5   Conclusions

The design and manufacture of a conical diffraction imaging spectrometer has been introduced in this paper. It is based on the off-plane Offner configuration, and is compact in its dimension. The convex blazed grating, the key component of the Offner imaging spectrometer, is manufactured by the holographic lithography - scan ion beam etching method in our lab. The alignment and measurement of the imaging spectrometer is also presented, and the measurement results show that this imaging spectrometer has excellent imaging quality.

Acknowledgements. The authors would like to acknowledge the fund support offered by A Project Funded by the Priority Academic Program (PAPD) and National Science Foundation of China (No. 61205188).

References

1. Bender, H.A., Mouroulis, P., et al.: Optical design, performance, and tolerancing of next-generation airborne imaging spectrometers. In: Proceedings of SPIE 7812, 98120P-1-12 (1997)
2. Bender, H.A., et al. Wide-field imaging spectrometer for the hyperspectral infrared imager (HyspIRI) mission. In: Proceedings of SPIE 9222, 92220E-1-8 (2014)
3. Offner, A.: New concepts in projection mask aligners. Opt. Eng. **14**(2), 130–132 (1975)
4. Das, N.C., Murty, M.V.R.K.: Flat field spectrograph using convex holographic diffraction grating and concave mirror. Pramäna J. Phys. **27**, 171–192 (1986)
5. Lucke, R.L.: Out-of-plane dispersion in an Offner spectrometer. Opt. Eng. **46**(7), 073004 (2007)
6. Prieto-Blanco, X., Montero-Orille, C., González-Núñez, H., et al.: Imaging with classical spherical diffraction gratings: the quadrature configuration. J. Opt. Soc. Am. A **26**(11), 2400–2409 (2009)
7. Prieto-Blanco, X., Montero-Orille, C., González-Núñez, H., et al.: The Offner imaging spectrometer in quadrature. Opt. Express **18**(12), 12756–12769 (2010)
8. Prieto-Blanco, X., González-Núñez, H., de la Fuente, R.: Off-plane anastigmatic imaging in Offner spectrometers. J. Opt. Soc. Am. A **28**(11), 2332–2339 (2011)
9. Pan, Q., Jin, Y., Zhao, Z., et al.: Design and analysis of conical diffraction imaging spectrometer with high spectral resolution and wide spectral dispersion range. In: Proceedings of SPIE 10021, 100211R (2016)
10. Liu, Q., Wu, J., Zhou, Y., et al.: The convex grating with high efficiency for hyperspectral remote sensing. In: Proceedings of SPIE 10156, 101561K (2016)
11. Parks, R.E., Kuhn, W.P.: Optical alignment using the point source microscope. In: Proceedings of SPIE 5877, 58770B (2005)

Motion Detection from Satellite Images Using FFT Based Gradient Correlation

Xichao Teng[(⊠)], Qifeng Yu, Jing Luo, Xiaolin Liu, and Xiaohu Zhang

National University of Defense Technology, Changsha, China
tengari@buaa.edu.cn

Abstract. In this paper, we study the problem of quantifying target motion at the Earth's surface. We leverage satellite images with different times to detect the movement of different targets. We focus on the motion detection of vehicle and exploit the near-simultaneous satellite images to assess the vehicle trajectories. We segment the target images into fixed size blocks and use FFT based gradient correlation to determine the displacements of each block. Then we reduce the block size and utilize an iterative multigrid image deformation method to calculate the global velocity field and improve the accuracy of motion detection. Compared to other correlation method, the FFT based gradient correlation is more accurate and time efficient, which can estimate translations, arbitrary rotations and scale factors. Moreover, the use of image gradient is able to capture the structure feature of salient image and make the correlation more robust. We do our experiments using images acquired by Planet Dove Satellites. The experiments show that our algorithm can quantify target motion robustly and efficiently. Our algorithm enhances the ability of time series satellite images to be used for motion detection.

Keywords: Motion detection · Small satellite images · FFT
Gradient correlation · Iterative multigrid · Image deformation

1 Introduction

Quantifying target motion at the Earth's surface using optical satellite images is one of the important application of satellite and remote sensing data. Accurate and reliable detection of target motion from satellite images is useful in the field of weather forecast, environment monitoring, traffic planning and military. By tracking the displacement of targets in an image time sequence, motions (vehicle motion, surface velocity field of cloud, water, glaciers and ice debris) can be quantified, and the results is evaluated for further application.

When satellite images were first exploited for motion detection, researchers manually selected corresponding targets and measured their displacement between different data acquisitions [1]. Later image matching algorithm based on correlation was used to perform motion detection automatically [2]. As the image matching method could effectively detect target motion with an image time series, many different image matching algorithms have been proposed to improve the robustness and accuracy of

© Springer Nature Switzerland AG 2018
H. P. Urbach and Q. Yu (Eds.): ISSOIA 2017, SPPHY 209, pp. 254–262, 2018.
https://doi.org/10.1007/978-3-319-96707-3_28

automatic motion detection. The main methods of image matching include normalized cross-correlation (NCC) [3]; Newton Raphson matching [4], phase correlation [5] and orientation correlation [6]. Heid and Kääb [7] compared and evaluated six different image matching methods for measuring glacier surface displacements, and they summarized that phase correlation and orientation correlation is the two most robust matching algorithms for glacier flow determination over large scales.

Our paper mainly focuses on the measurement of vehicle motion from optical satellite images with temporal resolution on the order of seconds to minutes. A few studies is related to the vehicle tracking with satellite data. Takasaki et al. [8] measured the speed vector of ship using JERS-1/OPS data. Reinartz et al. [9] perform traffic monitoring with airborne data. Pesaresi et al. [10] derived the velocity of moving targets using Quick Bird satellite data. Kääb and Leprince [11] provided general view on the exploitation of near-simultaneous images for Earth surface motion detection and describing methods that can be used to derive the vehicle trajectories in near real time data. Satellite images have seldom been used for vehicle tracking and most of these studies used spaceborne images acquired by Landsat, SPOT, ASTER or WorldView. However, with the current trends on small satellite development, a large number of near real time high resolution optical satellite images is now available for motion tracking, and their application to Earth surface process are little exploited. Xu [12] detected the vehicle motion in high resolution satellite video captured by Skysat and JL1H.

In this paper, we use high spatial and temporal resolution data acquired by planetscope (PS) satellites for vehicle tracking. The PS satellites can provide daily coverage and high spatial resolution monitoring of Earth from a satellite constellation currently numbering 132 orbital cameras. Each PS image consists of at least a red-green-blue (RGB) data with pixel size of 3–5 m. We exploit the PS image and demonstrate its application in motion detection. For motion detection algorithm, we use FFT based gradient correlation to estimate the displacements of moving targets and integrate iterative multigrid image deformation technique to improve the accuracy.

Our paper is organized as follows. In Sect. 2, our motion detection algorithm is described in detail. In Sect. 3, the experiments and results of motion detection are presented and discussed. Finally we conclude the paper in Sect. 4.

2 Methods

In this section, we will introduce our motion detection algorithm. Firstly, we divide satellite images into small blocks and leverage FFT based gradient correlation to estimate the relative motions between two blocks. Secondly, we reduce the size of blocks iteratively and update the relative motions estimated by FFT based gradient correlation until the size of blocks reaches the minimum. Finally, a clustering algorithm is used to get the displacements of moving targets.

2.1 FFT Based Gradient Correlation

Let I_1 and I_2 be two satellite images with different times, we denote the Fourier transform (FT) of image I by $\mathcal{F}\{I\}$ and inverse Fourier transform by $\mathcal{F}^{-1}\{I\}$. If there is an unknown translation $\mathbf{t} = \begin{bmatrix} t_x, t_y \end{bmatrix}^T$ between two images I_1 and I_2, we use standard phase correlation to estimate $\mathbf{t}$ from the normalized cross correlation function $PC(x, y)$, and the estimated translation $\hat{\mathbf{t}}$ can be represented as $\begin{bmatrix} \hat{t}_x, \hat{t}_y \end{bmatrix} = \arg \max_{(x,y)} \{PC(x, y)\}$, the normalized cross correlation function can be denoted by:

$$PC(x, y) = \mathcal{F}^{-1} \left\{ \frac{\mathcal{F}\{I_1\} \cdot \mathcal{F}^*\{I_2\}}{|\mathcal{F}\{I_1\} \cdot \mathcal{F}^*\{I_2\}|} \right\} \tag{1}$$

Where $*$ denotes the complex conjugate operator. Equation (1) is the normalized cross correlation function of standard phase correlation.

Considering the complexity of actual motion at the Earth's surface, there may be scaling and rotation between images besides translation. Although our motion detection algorithm mainly focuses on the vehicle tracking, we still expect that our algorithm is adaptive to other motions, for example, cloud motion. In our algorithm, we calculate normalized cross correlation function in the log-polar Fourier domain, which is also called Fourier Mellin transform (FMT). The FMT in form of the polar coordinate can be represented as follows:

$$M_f(k, v) = \frac{1}{2\pi} \int_0^\infty \int_0^{2\pi} f(r, \theta) r^{-iv} e^{-ik\theta} d\theta \frac{dr}{r} \tag{2}$$

where $f(r, \theta)$ is the function in polar coordinate (r, θ) and $M_f(k, v)$ is the FMT of $f(r, \theta)$. For two satellite images I_1 and I_2, if I_2 is a replica of I_1 with rotation angle $\theta_0 \in [0, 2\pi)$ and scaling factor $s > 0$, their FMTs are related by:

$$M_{I_2}(k, v) = M_{I_1}(k - \log s, v - \theta_0) \tag{3}$$

We can see that FT in the log-polar domain reduces the rotation and scaling to 2D translation which can be estimated by standard phase correlation. After estimating the rotation and scaling, we can calculate the remaining translation between two images using standard phase correlation.

As the size of vehicle in satellite images is often small, which means that the moving targets is not salient, we replace the images with their gradients for cross correlation. Gradient correlation (GC) combines the magnitude and orientation of image gradients, the gradient image $G(x, y)$ is denoted by:

$$G(x, y) = \frac{\partial I}{\partial x} + i \frac{\partial I}{\partial y} \tag{4}$$

Orientation correlation (OC) only considers orientation information, while GC uses the gradient magnitude and is more stable [13]. Our algorithm uses GC for images with weak contrast.

Our FFT based gradient correlation uses phase correlation twice: once in the log-polar Fourier domain to estimate the scaling and rotation and once in the spatial domain to estimate the translation. And in two steps, we extract image complex gradients for FT. Figure 1 illustrates the process of FFT based gradient correlation algorithm:

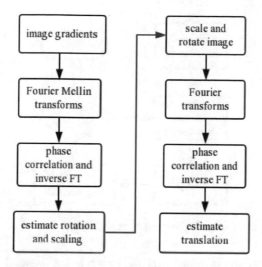

Fig. 1. Flow chart of FFT based gradient correlation algorithm

2.2 Iterative Multigrid Image Deformation

Scarano and Riethmuller [14] proposed iterative multigrid image deformation (IMID) technique for particle image velocimetry (PIV) which combined iterative prediction, image deformation and progressive grid refinement. In our paper, we use this technique to improve the accuracy of vehicle tracking. In our algorithm, we do not detect vehicle in the satellite images by feature extraction or learning methods and we search the moving targets in the whole image, so we use IMID technique to identify the displacement of every pixel and threshold pixels with incorrect displacements (too big or too small).

The main steps of iterative multigrid image deformation method are as follows:

Step 1. Divide the image into a grid and estimate the displacements of grid nodes by FFT based gradient correlation method.
Step 2. Obtain the displacement of each pixel in the image by interpolation according to the displacements of grid nodes, and get the displacement field of the whole image.
Step 3. Deform the image according to the displacement field using interpolation method.
Step 4. Refine the grid and estimate small displacements to get the displacement field of the refined grid.
Step 5. Perform steps 2–5 iteratively until the number of iterations is reached. Figure 2 is the flow chart of IMID method.

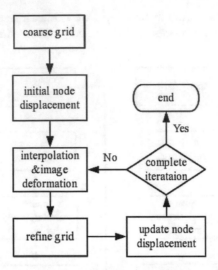

Fig. 2. Flow chart of IMID algorithm

Our motion detection algorithm integrate FFT based gradient correlation method and IMID technique to derive the velocity field of the satellite image. As the result of IMID method is the displacement of every pixel, we need to identify velocity vectors which correspond to vehicle motions. In the next section, we will introduce the clustering method for vehicle tracking.

2.3   Clustering

As our algorithm does not locate the moving targets such as ships, cars and airplanes and we estimate the displacement of every pixel between satellite images with different times, the displacements of moving targets need to be identified for vehicle tracking. In our algorithm, we use clustering method to find the displacements which belong to the vehicle. For satellite images with moving targets such as ships, cars and airplanes, we can determine the range of target displacements according to the approximate speed of vehicle and time interval. After the determination of displacement range, we use the range as thresholds to remove incorrect displacements. For the remaining displacements, we cluster them according to the similarity of magnitude, orientation and spatial position and we identify clustering centers as locations of moving targets.

Figure 3 show the clustering result of vehicle motion. As we can see, the clustering algorithm identify the displacement of the car automatically.

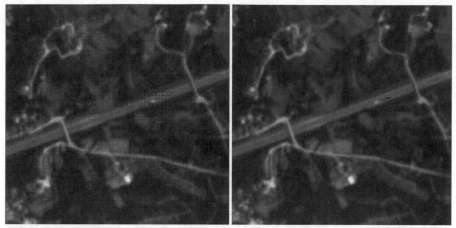

(a) displacement field obtained by   (b) vehicle displacement by clustering
    IMID method

Fig. 3. Clustering result which identifies the displacement of a car on a highway in Changsha, Hunan, China. The image is a section of PS satellite image on 27 March 2017 with a time lag of 2 s.

3 Experiments

We use PS satellite images to verify the performance of our motion detection algorithm. In our experiments, we detect motions of airplanes, cars and ships.

In Fig. 4, we show the airplane speed on the Hong Kong International Airport, China, from PS satellite images of 22 July 2017 and a time lag of 100 s. In Fig. 4(c) we fuse two images to show the vehicle motion. The minimum size of image blocks for correlation is 32×32. According the displacement and time lag, we estimate the airplane speed of 1.2 m/s.

In Fig. 5, we show the ship speed in the sea area near Changshan Island, China, from PS satellite images of 18 May 2017 and a time lag of 31 s. In Fig. 5(c) we fuse two images to show the vehicle motion. The minimum size of image blocks for correlation is 16×16. According the displacement and time lag, we estimate the ship speed of 1.9 m/s.

For the estimation of ship speed, surface waves in the sea would disturb the clustering of displacements as our algorithm does not detect ships using feature extraction or learning methods, we will improve the robustness of our algorithm against the interference of sea waves or illumination changes.

(a) The PS satellite image over Hong Kong International Airport

(b) The PS satellite image over Hong Kong International Airport

(c) Displacement vector estimated by our motin detection algorithm

Fig. 4. Result of our motion detection algorithm for the airplane speed on the Hong Kong International Airport, China.

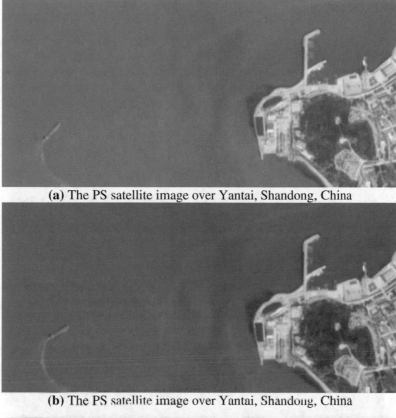

(a) The PS satellite image over Yantai, Shandong, China

(b) The PS satellite image over Yantai, Shandong, China

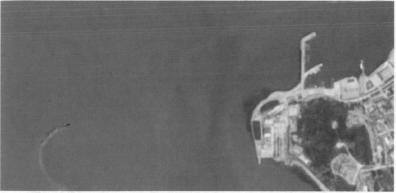

(c) Displacement vector estimated by our motin detection algorithm

Fig. 5. Result of our motion detection algorithm for the ship speed in the Bohai Sea near Changshan Island in Yantai, Shandong, China.

4 Conclusion

In this paper, we propose an algorithm to quantify target motion at the Earth's surface, and we exploit the feasibility of PS satellite images for vehicle tracking. For vehicle tracking with satellite images, our algorithm leverage gradient correlation for the robust matching of targets with weak visual contrast. In addition, our method dose not need prior information to identify the target's location and is more feasible compared with other vehicle tracking processes.

References

1. Lucchitta, B.K., Ferguson, H.M.: Antarctica: measuring glacier velocity from satellite images. Science **234**(4780), 1105–1108 (1986)
2. Bindschadler, R.A., Scambos, T.A.: Satellite-image-derived velocity field of an Antarctic ice stream. Science **252**(5003), 242–246 (1991)
3. Robert, B., et al.: Surface velocity and mass balance of ice streams D and E, West Antarctica. J. Glaciol. **42**(142), 461–475 (1996)
4. Thinzar, C.: Detecting the storm movement by sub pixel registration approach of Newton Raphson method. Int. J. e-Educ. e-Bus. e-Manag. e-Learn. **4**(1), 28–31 (2014)
5. Puymbroeck, V.: Monitoring earth surface dynamics with optical imagery. Geology **89**(1), 1–12 (2008)
6. Fitch, A.J., et al.: Orientation correlation. In: British Machine Vision Conference 2002, Cardiff, UK, 2–5 September 2002, pp. 133–142. DBLP (2002)
7. Heid, T., Kääb, A.: Evaluation of existing image matching methods for deriving glacier surface displacements globally from optical satellite imagery. Remote Sens. Environ. **118**, 339–355 (2012)
8. Takasaki, K., Sugimura, T., Tanaka, S.: Speed vector measurement of moving objects using JERS-1/OPS data. In: Better Understanding of Earth Environment, International IEEE Geoscience and Remote Sensing Symposium, 1993, vol. 2, pp. 476–478 (1993)
9. Reinartz, P., et al.: Traffic monitoring with serial images from airborne cameras. J. Photogramm. Remote Sens. **61**(3), 149–158 (2006)
10. Pesaresi, M., Gutjahr, K.H., Pagot, E.: Estimating the velocity and direction of moving targets using a single optical VHR satellite sensor image. Int. J. Remote Sens. **29**(4), 1221–1228 (2008)
11. Kääb, A., Leprince, S.: Motion detection using near-simultaneous satellite acquisitions. Remote Sens. Environ. **154**, 164–179 (2014)
12. Xu, A., Wu, J., Zhang, G., Pan, S., Wang, T., et al.: Motion detection in satellite video. J Remote Sens. GIS **6**, 194 (2017)
13. Tzimiropoulos, G., et al.: Robust FFT-based scale-invariant image registration with image gradients. IEEE Trans. Pattern Anal. Mach. Intell. **32**(10), 1899–1906 (2010)
14. Scarano, F., Riethmuller, M.L.: Advances in iterative multigrid PIV image processing. Exp. Fluids **29**(1), S051–S060 (2000)

A Study on Monitoring Method of High-Resolution Remote Sensing in Cloudy and Rainy Regions

Shanwei Liu, Bin Yu, Naixin Zhang, Jinxia Zang, Jiaxin Wan,
and Jianhua Wan[✉]

School of Geosciences, China University
of Petroleum (East China), Qingdao, China
wjh66310@163.com

Abstract. Remote Sensing technology has become an effective method to gain basic geographic information of large scale in coastal zones. But the areas along the Silk Road mostly located in the tropics, its cloudy and rainy weather acquires a high-resolution optical image. We used high-resolution full-polarimetric SAR and medium resolution optical image together instead of high-resolution optical image to carry out the classification of this area. We analyzed the polarization component characteristics of the ground object via full polarization SAR and evaluated the feasibility of classification. The cloud in the medium resolution optical image was removed based on high-resolution SAR, and then the optical image fused with SAR, which has the same effect as the high-resolution optical image.

Keywords: Cloudy and rainy regions · High-resolution remote sensing
Classification · Maritime Silk Road

1 Introduction

The 21st Century Maritime Silk Road is a new trade route for China to connect the world to the changing situations of global politics and trade patterns. To ensure the security of the transportation corridor, it is necessary to master the basic geographic information of regions along the Silk Road, which is also of great value for coastal countries to prevent and mitigate disasters and develop their economies.

Remote sensing technology has become an important means for the rapid and efficient acquisition of geographic information of large-scale base in coastal zones. However, as for high-resolution optical satellite image, it features small width, long revisit period, and it is subject to the limit of the meteorological conditions. Therefore, in cloudy and rainy tropical regions, it can reserve fewer data. And it is difficult to obtain qualified images in a short time even using programming. SAR is not disturbed by cloud fog, easy to get information quickly, and the high-resolution SAR has reached the sub-meter resolution; The full polarimetric SAR possesses more multi-polarization information than traditional single polarization and dual polarization, which significantly improves the classification ability and has the feasibility of replacing the high-resolution

© Springer Nature Switzerland AG 2018
H. P. Urbach and Q. Yu (Eds.): ISSOIA 2017, SPPHY 209, pp. 263–271, 2018.
https://doi.org/10.1007/978-3-319-96707-3_29

optical image. However, without spectral information of SAR, the accuracy of the classification of ground objects may be reduced. If combining the moderate-resolution optical image, the limitations of SAR lack of spectral information may be compensated to some extent. Being different from the selective shooting mode of high-resolution optical satellite, the medium resolution optical satellite has been in a state of the shoot, and it features big width, short revisit period. Therefore, its time window of good weather is still available in cloudy and rainy areas, and images-taken can be guaranteed.

Based on the above situation, the paper utilized high polarization SAR, with the combination of the medium resolution optical image in the cloudy and rainy region featuring classification method research, to achieve high classification level of optical images.

2 Feasibility Evaluation of High-Resolution Fully Polarimetric SAR Classification

The fully polarimetric SAR has four polarization modes, containing more polarization information than monopolar and dual polarized SAR. The polarization decomposition technology utilizes the additional information implied in the cross-polarization items to extract more polarization characteristics. Polarization decomposition methods mainly include two categories; one is noncoherent decomposition [1–3], the other is coherent decomposition [1, 4, 5].

In this paper, the Radarsat-2 images (the distance resolution is 12 m, the azimuth resolution is 8 m) which taken in July 2009, were used to evaluate the classification feasibility of Polarization component. Firstly, we preprocessed the original date, including geometric correction, denoising, generating covariance matrix. Then we used the Freeman Decomposition method to extract polarimetric features including Surface scattering, Volume scattering, Oriented Dihedral Scattering. Finally, according to the importance of research and distribution characteristics in the study area, we selected nine common objects for analysis, including Reed, Cultivated Land, Suaeda, Tamarix Chinensis Lour, Prairie Cordgrass, Locust Forest, Tidal Flat, Water, and Culture Pond. We extracted Backscattering coefficients of nine kinds of objects, and calculated average backscattering coefficients. The various response curves are shown in Fig. 1.

(1) In the even scattering channel (Fig. 1a), there are few intersections between different ground objects. Even generation scattering is limited by special structural factors, so the overall response value of the objects is very low, presenting a large area of black in the image. It included the arable land, Tamarix, Spartina, Robinia, tidal flat, water, sea aquaculture pond, with a majority of things confused together, it is difficult to distinguish precisely. But the reed and the base of the two baskets are opposite, their response was bright, and the response between the two is obvious, so the reed and the sorghum can be determined based on even scattered channel information.

(2) In the body scattering channel (Fig. 1b), there are three types of clustering phenomena, in which the response curve of water and rice grass is low, and the response curve of tidal flat and pond is located in the middle, while the response of reed, cultivated land is high. In response to the high terrain species in the

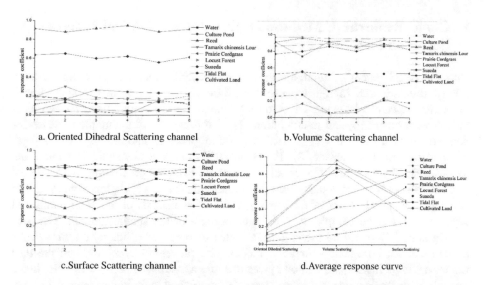

Fig. 1. The response coefficient curve of different scattering channels

biological form are lush vegetation, in which the scattering response of *Robinia pseudoacacia* is stable and at the highest level. These three kinds of clustering categories can be distinguished, but the categories of objects are confusing, and using the body scattering channel information is not enough to judge.

(3) The surface scattering channel (Fig. 1c) is the most balanced component of the scattering effect distribution. There is no large area black spot and no serious confounding phenomenon. The contours of each category in the image are clearly visible and the most comprehensive scattering components of the original feature are preserved. The response coefficient of reed is in the low level in the surface scattering channel, and its response in the even scattering channel has obvious contrast characteristics, which can be used to improve the accuracy of the classification of such objects.

(4) According to the analysis of the three-component average response curve (Fig. 1d), the overall response of the even scattering channel is the weakest, and the lowest scattering value of the other objects except the reed appears in the even scattering channel, and most of the whole image is black. The integrated response of the bulk scattering channel is significantly higher than that of the even scattering, and the surface scattering is slightly higher than the bulk scattering. From the point of view of different types of species, the response of the two species in the body scattering channel was the most significant, which was closely related to their own lush vegetation structure. Reed, The strongest scattering mechanism of water and seawater pond is surface scattering.

3 Cloud Removal of Medium Optical Images Optical Images Based on SAR

With the development of multi-spectral, hyperspectral and microwave sensor technology, domestic and foreign scholars have carried out some remote sensing image restoration research, in which involved algorithms can be summarized as two categories: replacement [6] and interpolation [7]. The replacement technique replaces the cloud pixel value (DN) with the same position pixel in another or other images obtained by the same sensor [8]. Its advantage is that the algorithm is simple and operable. But there are many potential error sources in the preprocessing chain, and the time differences between images are often neglected. The interpolation technique is to interpolate the pixel values of the missing cloud with analysis of a certain number of pixels, which can be completed in the time domain or the spatial domain. The time domain is mainly based on the composite index of time series image, the time series outliers and the numerical estimation model. The process is complex, and the data requirements are high. Interpolation in spatial domain assumes that the variability in the image is spatially dependent, and the interpolation method is limited to the sensor type, ground cover type, cloud type, and so on, and the versatility is poor.

Considering the universality of the image reconstruction algorithm, this article integrates the replacement and interpolation techniques to achieve cloud elimination based on the optimal pixel retrieval matching from intraclass pixel similarity and cloud image reconstruction. The experimental data are optical images of the fifth, fourth, third, and second wave band obtained by NASA Earth Observation Satellite Landsat8 OLI sensors on August 8, 2015, and SAR data of the full polarization fine pattern from Canadian radars at-2 satellite on September 16, 2015.

The thick cloud detection and extraction and cloud shadow detection and extraction are carried out based on optical images, and then cloud noise is removed. Then, the Radarsat-2 image of the same period is fused, and the color distance of the pixel is introduced to the fusion reconstruction algorithm as the weight, and the spatial distance between pixels as weight improves the traditional Euclidean distance W_{ED}, the formula is:

$$W_{ED} = \sqrt{\sum_{i=1}^{k} \left(S_{x,y,k} - S_{i,j,k}\right)} * \sqrt{(x-i)^2 + (y-j)^2} \qquad (1)$$

Among them, K is the number of video bands, x, y, i and j is corresponding pixel column Numbers. The steps are to search for reconstructed pixels, and then to replace cloud noise pixels to generate new remote sensing images. The overall process is shown in Fig. 2.

Based on the similarity of the spectral similarity of the ground objects and the corresponding relationship between the data of Radarsat-2 polarimetric SAR data, the new non-cloud multi-spectral data is reconstructed by the algorithm of this paper, as shown in Fig. 3(b). Based on pixel replacement similar, cloud elimination algorithm removes cloud noise well, retains the original multispectral image color information, and makes the reconstructed image color contrast and fusion more apparent, the texture information more abundant. From the detailed comparison of Fig. 3(a) and (b), we can

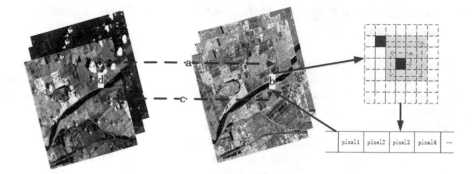

Fig. 2. The flow of data fusion reconstruction

see: although the reconstructed images had minor patches, the clouded and shadowed parts of the image was effectively repaired, which improves the visual effect. As the algorithm of this article is based on pixel units, the edge smoothing is not processed after extracting the cloud and shadow from multispectral images, and proper edge processing methods should be used to improve the fusion effect in the next step.

Fig. 3. Landsat8 false-color image (a), Cloud-free image (b) and image details (c, d)

4   Classification of High-Resolution SAR and Medium Optical Image Fusion

SAR and optical images have their respective advantages, and the studies of high-resolution SAR and the medium, the optical image is increasing. In the field of SAR and optical image pixel fusion [9–12] have developed many new fusion ways, such as PCA, HSV, wavelet transform, and achieved a better fusion effect.

HSV fusion method can effectively superimpose the spectral and spatial characteristics of the image. But when SAR and multispectral image fuse, the significant difference between multi-source data brightness will lead to fusion image color distortion. So it is often used in conjunction with other methods. The wavelet transform is a multi-scale decomposition of the image. It can fuse high spatial resolution information into low spatial resolution images of the various bands. So the image color information is reserved, and the structure and details are highlighted. In this paper, we use a correlation-weighted HSV transformation and high-resolution SAR in wavelet transformation method and mid-grade multi-spectral image fusion algorithm to fuse images. The fusion image is shown in Fig. 4: the edges of farmland, road and so on are more obvious, and their colors and details are expressed better, and image interpretation ability is greatly improved. A solution for the simulation of the high-resolution optical image is provided.

Fig. 4. SAR-VV polarimetric image (a), fusion image of non-cloud (b) and image details (c, d)

Using recognition software, the image is multi-scale structurally segmented. Its the basic unit is the image. The nearest classification method is used to classify the images, and then the accuracy of the classification result is evaluated and analyzed. The classification results are shown in Fig. 5 below.

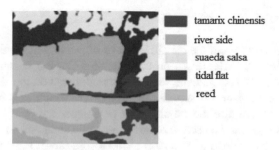

Fig. 5. The classification image of fusion result

According to the survey data of the Yellow River estuary area, 20 samples were randomly selected for each feature (Table 1). The accuracy of the classification was calculated with the confounding matrix.

Table 1. The classification accuracy of different land cover types

Land cover types	Producer accuracy	User accuracy
tamarix chinensis	78.98	77.66
riverside	87.99	97.79
suaeda salsa	85.65	97.98
tidal flat	93.44	74.35
reed	84.19	87.63
Overall accuracy (%)	88.17	
Kappa coefficient	0.84	

(1) The object-oriented classification of fusion image based on multi-scale segmentation can effectively avoid the influence of salt and pepper noise in the classification result of traditional pixel-based classification. At the same time, the structural features of objects, spectral characteristics, etc. are applied to the classification, combined with the spatial relationship of the target object to reduce the spatial heterogeneity and achieve good classification effect.

(2) The overall accuracy of the fusion image classification is 88.17%, and the Kappa coefficient is 0.84, which means that the classification result is more authentic and the desired effect is achieved. The optimal accuracy of the tidal flats is 93.44%, the highest. The accuracy of river water surface and Suaeda in the saline land are relatively same high that shows good separability. In general, the information fusion of vein and spectrum achieves multi-faceted identification of objects. Using object classification, we can effectively reduce the cases of "same-spectrum in different objects" and "same object with different spectrums" and achieve better classification accuracy.

5   Conclusion

(1) It is feasible to carry out the classification of objects by using the high-resolution and full polarized SAR. Based on the Freeman–Durden decomposition method, the response values of the three scattering components on the channel of different scattering mechanism are significantly different, which can be used as useful information to optimize the classification of objects.
(2) The cloud image can be used to remove the cloud image from the optical image. Through the detection and extraction of the resolution of the optical image of the cloud and shadow occlusion area, based on SAR images to achieve optical image cloud area repair, to achieve the ideal image to cloud effect.
(3) Based on SAR and optical image fusion data can achieve better classification effect. Joint HSV transform and wavelet transform to achieve optical and SAR image fusion enhancement. The object-oriented classification is used to classify the fusion image, and the classification effect is achieved.

Acknowledgments.   This research has been supported by National Key R&D Program of China (2017YFC1405600), National Natural Science Foundation of China (41776182) and Shandong Provincial Natural Science Foundation of China (ZR2016DM16).

References

1. Cloud, S.R., Pottier, E.: A review of target decomposition theorems in radar polarimetry. IEEE Trans. Geosci. Remote Sens. **34**(2), 498–518 (1996)
2. Freeman, A., Durden, S.L.: A three-component scattering model for polarimetric SAR data. IEEE Trans. Geosci. Remote Sens. **36**(3), 963–973 (1998)
3. Yamaguchi, Y., Moriyama, T., Ishido, M., Yamada, H.: Four-component scattering model for polarimetric SAR image decomposition. IEEE Trans. Geosci. Remote Sens. **43**(8), 1699–1726 (2005)
4. Kroger, E.: New decomposition of the radar target scattering matrix. Electron. Lett. **26**(18), 1525–1527 (1990)
5. Cameron, W.L., Leung, L.K.: Feature motivated polarization scattering matrix decomposition. In: IEEE International Radar Conference, Arling, VA, pp. 549–557 (1990)
6. Lu, D.: Detection and substitution of clouds/hazes and their cast shadows on IKONOS images. Int. J. Remote Sens. **28**(18), 4027–4035 (2007)
7. Yao, C., Jinliang, W.: The methods for removing the effects of cloud cover in remote sensing images. Remote Sens. Land Resour. **18**(1), 61–65 (2006)
8. Pyongsop, R., Zhangbao, M., Qingwen, Q., et al.: Cloud and shadow removal from Landsat TM data. J. Remote Sens. **14**(3), 534–545 (2010)
9. Pal, S.K., Majumdar, T.J., Bhattacharya, A.K.: ERS-2 SAR and IRS-1C LISS III data fusion: a PCA approach to improve remote sensing based geological interpretation. ISPRS J. Photogramm. Remote Sens. **61**, 281–297 (2007)

10. Amolins, K., Zhang, Y., Dare, P.: Wavelet-based image fusion technique—an introduction, review, and comparison. ISPRS J. Photogramm. Remote Sens. **62**(4), 249–263 (2007)
11. He, C., Liu, Q., Li, H.: Multimodal medical image fusion based on IHS and PCA. Procedia Eng. **7**, 280–285 (2010)
12. Zhang, Y., Gong, L., Li, H.: Merge of laser radar image and optical image. In: International Conference on Earth Science and Remote Sensing, vol. 30, pp. 304–309 (2012)

Author Index

© Springer Nature Switzerland AG 2018
H. P. Urbach and Q. Yu (Eds.): ISSOIA 2017, SPPHY 209, pp. 273–274, 2018.
https://doi.org/10.1007/978-3-319-96707-3

Printed in the United States
By Bookmasters